SpringerBriefs in Geography

For further volumes:
http://www.springer.com/series/10050

Yen-Chiang Chang

Ocean Governance

A Way Forward

Springer

Yen-Chiang Chang
School of Law
Shandong University
Hongjialou 5
250100 Jinan, Shandong
People's Republic of China
e-mail: ycchang@sdu.edu.cn

ISSN 2211-4165　　　　　　　　　　　e-ISSN 2211-4173
ISBN 978-94-007-2761-8　　　　　　　e-ISBN 978-94-007-2762-5
DOI 10.1007/978-94-007-2762-5
Springer Dordrecht Heidelberg London New York

Library of Congress Control Number: 2011940781

© The Author(s) 2012
No part of this work may be reproduced, stored in a retrieval system, or transmitted in any form or by any means, electronic, mechanical, photocopying, microfilming, recording or otherwise, without written permission from the Publisher, with the exception of any material supplied specifically for the purpose of being entered and executed on a computer system, for exclusive use by the purchaser of the work.

Cover design: Deblik, Berlin

Printed on acid-free paper

Springer is part of Springer Science+Business Media (www.springer.com)

Preface

Ocean—a source of life, has been overused and heavily polluted. While the traditional approaches may not be able to solve the complexity of the ocean governance issues, there is a need to adopt a new way of thinking in order to deal with the current problems. This book emphasises the importance of law and policy while generating ocean governance initiatives. "Good Ocean Governance" as a new way of thinking, needs to be supported by legislation and decision makers. It is, therefore, necessary to examine whether the idea of good ocean governance exists within the international legal system and is subsequently subject to transfer into the domestic law. During this transaction process, a number of examples provided from the United States, Canada and Australia demonstrate the broad picture. The picture would not be entirely clear without discussions on the governance of marine resources, ship recycling and marine pollution, the impact of maritime clusters as well as social and culture impact of ports. The ultimate aim of this book is to tease out more new ideas and talks on ocean governance issues.

Contents

Chapter 1
Getting Into a New Era: Good Ocean Governance

Abstract This chapter proposes eight elements which are the more commonly accepted as elements which contribute to the concept of good governance, these being, the *rule of law, participatory, transparency, consensus based decision making, accountable, equitable and inclusive, responsive* and *coherent*. The chapter also provides examples from international treaty and State practice for each element of good governance. The aim of this exercise is to distinguish each individual element of good governance for the purposes of establishing criteria for a further examination of practice. The outcome also provides an illustration as to what the elements of good governance may look like in international treaty and State practice. It can be concluded that each element of good governance is, to some extent, supported by international treaty and State practice.

Keywords Good ocean governance · Sustainable development · The rule of law · Transparency

1.1 Introduction

The current policy development in relation to oceans governance at international level is that existing mechanisms provide only sectoral governance structures. There are no clear mechanisms or policy approaches in place to foster cooperation and coordination in a way that could comprehensively and effectively address the conservation of marine ecosystems.[1] There are, rather, various bodies, which have specific objectives of their own.

The World Bank, for example, is considered to be a powerful instrument in fostering sustainable use of marine ecosystems, due to its role in international cooperation and global partnerships. While relevant marine ecosystem components

---

[1] UN, *Report of Secretary-General*, 62nd Session 2007, A/62/66/Add.2, pp. 75.

Y.-C. Chang, *Ocean Governance*, SpringerBriefs in Geography,
DOI: 10.1007/978-94-007-2762-5_1, © The Author(s) 2012

are included in the design of the World Bank's projects, it is the fact that the latter is mindful of the difficulties in implementing good ocean governance policy at the national, regional and global levels.[2]

At the European level, the current shortcomings of a sectoral based approach have been identified[3] and, as a result, a more holistic good ocean governance approach is being encouraged.[4] As the *EU Maritime Green Paper* states, "Principles of good governance suggest the need for a European maritime policy that embraces all aspects of the oceans and seas."[5] The European Commission further explains that good governance and an integrated approach could help to move towards a more overarching strategy, which would more effectively unite the present sectoral policies for maritime activities and environmental policy.[6] Unfortunately, the European maritime policy does not seem to indicate clearly how good ocean governance might be achieved.

The current policy proposals in the UK also do not set out a clear approach as to how good ocean governance can be achieved. In 2002, the UK Government published *Safeguarding Our Seas*,[7] which sets out its vision for "clean, healthy, safe, productive and biologically diverse oceans and seas." In particular, it emphasizes the importance of stakeholder involvement and good fisheries governance.[8]

In Scotland, *Sea the Opportunity*[9] states that one of the key elements of maritime strategy is "promoting good governance."[10] *A Sea Change—A Marine Bill White Paper*[11] referred to the aforesaid proposal and it also emphasizes the importance of "promoting good governance" as an overarching approach to achieving the UK's sustainable development goal.[12] Unfortunately, none of the

---

[2] UN, *Report of Secretary-General*, 61st Session 2006, A/61/63, pp. 56–57.

[3] Commission of the European Communities, *Towards a future Maritime Policy for the Union: A European Vision for the Oceans and Seas*, COM (2006) 275 final, Vol. I, pp. 4.

[4] Commission of the European Communities, *An Integrated Maritime Policy for the European Union*, COM (2007) 575 final, pp. 3–5; Commission of the European Communities, *Commission Staff Working Document*, SEC (2007) 1280, pp. 2.

[5] Commission of the European Communities, *Green Paper—Towards a future Maritime Policy for the Union: A European Vision for the Oceans and Seas*, COM (2006) 275 final, Vol. II-ANNEX, pp. 5.

[6] Commission of the European Communities, *Commission Staff Working Document*, SEC (2007) 1278, pp. 10.

[7] DEFRA, *Safeguarding Our Seas—A Strategy for the Conservation and Sustainable Development of our Marine Environment*, 2002.

[8] *Ibid*, pp. 57.

[9] Scottish Executive, *Sea the Opportunity—A Strategy for the Long Term Sustainability of Scotland's Coasts and Seas*, August 2005.

[10] *Ibid*, pp. 6.

[11] DEFRA, *A Sea Change—A Marine Bill White Paper*, March 2007, Cm 7047.

[12] *Ibid*, pp. 8.

aforementioned proposals indicate clearly what elements contribute to 'good governance'.

The central issue to be considered throughout this chapter is how best to govern the marine environment. 'Sustainable development', is of primary importance when considering marine environmental protection. The concept of sustainable development, however, has been criticized as being too vague and imprecise in content[13] and as a result, a clear and workable indicator is essential if progress is to be achieved.

'Good governance', is considered to be a positive and constructive element of sustainable development.[14] It is an open decision-making process which should involve public participation, the release of environmental information and access to environmental justice.[15] While there is a great deal of literature which considers the issue of 'good governance', what constitutes the elements of good governance is still the subject of debate. This chapter will, therefore, review literature in relation to good governance, in order to establish the elements which are required in order to constitute good governance. The paper will then draw conclusions as to which elements are essential to the concept of good governance.

Succinctly, it is proposed that there are eight elements which appear to be accepted as being the essential elements of good governance, which are, rule of law, participatory, transparency, consensus based decision making, equity and inclusiveness and responsiveness and coherence. These elements of good governance are considered, to some extent, to overlap.[16] Examples from international treaty and State practice will be provided to support each element of good governance. The aim is to distinguish each individual element of good governance for the purposes of establishing criteria for a further examination of practice.

---

[13] Todd B. Adams, "Is there a Legal Future for Sustainable Development in Global Warming? Justice, Economics, and Protecting the Environment", The Georgetown International Environmental Law Review (2003–2004): 16; Nancy Nelson, "Sustainable Development", Colorado Journal of International Environmental Law and Policy (1997): 59–61; Lynda M. Warren, "Sustainable Development and Governance", Environmental Law Review (2003): 79–81.

[14] Konrad Ginther and Paul J. I. M. de Waart, "Sustainable Development as a Matter of Good Governance: An Introductory View", in Konrad Ginther, Erik Denters and Paul J. I. M. de Waart (ed), *Sustainable Development and Good Governance*, (London: Martinus Nijhoff Publishers, 1995), pp. 9.

[15] See Premable of the Convention on Access to Information, Public Participation in Decision-Making and Access to Justice in Environmental Matters (Aarhus) 25 June 1998, came into force on 30th October 2001, 38 ILM 517 (1999).

[16] Organisation for Economic Co-operation and Development, *Improving Policy Coherence and Integration for Sustainable Development: A Checklist* (2002), pp. 2; Francis N. Botchway, "Good Governance: The Old, The New, The Principle, and The Elements", Florida Journal of International Law (2001) no.13: 180–183.

1.2 The Elements of Good Governance

This section aims to explore the constituent elements which contribute to 'good governance'. Before tackling 'good governance', there is a need to explore the term 'governance'. There is no single definition of 'governance' and different approaches may be taken when considering its implication. Although the word 'governance' has to be generally understood as referring to actions dominated by government,[17] it is also commonly used in relation to private companies such as corporate governance.

The United Nations Development Programme (UNDP) defines governance as "the exercise of economic, political and administrative authority to manage a country's affairs at all levels. It comprises the mechanisms, processes and institutions through which citizens and groups articulate their interests, exercise their legal rights, meet their obligations and mediate their differences."[18] Similarly, the European Commission interprets governance as something that "concerns the state's ability to serve the citizens. It refers to the rules, processes, and behaviours by which interests are articulated, resources are managed, and power is exercised in society."[19]

The World Bank captures the economic dimensions of governance as being "the manner in which power is exercised in the management of a country's economic and social resources for development"[20] The Asian Development Bank in considering its operation and defining the term governance as "the manner in which power is exercised in the management of a country's economic and social resources for development."[21]

Other writers interpret governance as "the way in which stakeholders interact with each other in order to influence the outcomes of public policies."[22] Rotberg perceives governance as "the term used to describe the tension-filled interaction between citizens and their rulers and various means by which governments can either help or hinder their constituents' ability to achieve satisfaction and material prosperity."[23] Rhodes understands governance as "a change in the meaning of government, referring to a new process of governing; or a changed condition of

[17] See *Concise Dictionary* (Harper Collins Publishers, 2001), p. 623; see also *Longman Dictionary of the English Language* (Viking, 1995), p. 682.

[18] UNDP, *Governance for Sustainable Human Development—A UNDP Policy Document*, January 1997, available from: http://mirror.undp.org/magnet/policy/ last visited date: 21/12/2007.

[19] The European Commission, *Communication on Governance and Development*, October 2003, COM (03) 615.

[20] The World Bank, *Governance and* Development, Washington, 1992, p. 1; UNDP, *Governance Indicators: A User's Guide*, p. 3, available from: http://www.undp.org/oslocentre/docs04/UserGuide.pdf last visited date: 21/12/2007.

[21] Asian Development Bank, *Governance: Sound Development Management*, August 1999, p. 3.

[22] Tony Bovaird and Elke Löffler, "Evaluating the Quality of Public Governance: Indicators, Models and Methodologies", International Review of Administrative Sciences (2003), 69 no. 3: 316.

[23] Robert I. Rotberg, "Strengthening Governance: Ranking Countries Would Help", Washington Quarterly (Winter 2004–5), 28:1, p. 71.

ordered rule; or the new method by which society is governed."[24] Kjær promotes governance as "a mechanism that will help aid projects to promote economic growth and reduction of poverty, by becoming transparent, accountable and by following the rule of law."[25] Pierre states that governance "is a rejection of the New Right's claim that market mechanisms should substitute for political steering whenever possible."[26]

Based on the aforesaid, governance can be perceived as being the relationship between a society and its government.[27] The concept of governance, therefore, encompasses not only law and the public authorities but also relates to government policies and its implementations.[28] It is now seen as the key that locks policy making and policy implementation together.[29] In an environmental connotation covering both human actions and environmental consequences, it comprises both "a series of technical and generally well defined measures governing physical interactions between human activities and marine environment; and [...] a general management dimension encompassing co-ordination of technical management measures, organisational decision-making, policy and strategic planning aspects."[30]

The literature in relation to the development of the concept of good governance across time is also reviewed. The World Bank, in particular, has put a great deal of effort into developing indicators for evaluating the utility of development assistance and endorsing 'good governance' as a core element of its development strategy.[31] It has used an aggregate approach to identify three governance indicators, these being *government effectiveness*, *rule of law* and *graft*.[32] This

---

[24] R. A. W. Rhodes, "The New Governance: Governing without Government", Political Studies (1996): 652–3.

[25] A. M. Kjær, *Governance* (Cambridge: Polity Press, 2004), Introduction.

[26] Jon Pierre ed., *Debating Governance: Authority, Steering, and Governance* (New York: Oxford University Press, 2000), pp 264.

[27] Peter Rogers and Alan W Hall, *Effective Water Governance*, Global Water Partnership Technical Committee, The Background Papers, No. 7, February 2003, p. 4.

[28] Marie Besançon, *Good Governance Ranking: The Art of Measurement*, World Peace Foundation Report, Number 36, 2003, p. 1, 7 and 8; see also Robert I. Rotberg, supra note 23, p. 71; Peter Rogers and Alan W Hall, ibid, p. 4.

[29] Andrew Allan and Patricia Wouters, 'What Role for Water Law in the Emerging "Good Governance" Debate?', available from http:// www.dundee.ac.uk/water/Documents/Conferences/ 2004/woutersGovernancearticleforAWRA.pdf. last visited date: 18/03/2008.

[30] Hance D. Smith and Jonathan S. Potts, 'People of the Sea—The British Maritime World', in Hance D. Smith and Jonathan S. Potts (ed), *Managing Britain's Marine and Coastal Environment—Toward a Sustainable Future* (London: Routledge, 2005), p. 14.

[31] Patrick Molutsi, "Tracking Progress in Democracy and Governance Around the World: Lessons and Methods", available from: http://unpan1.un.org/intradoc/groups/public/documents/ un/unpan005784.pdf last visited date: 22/12/2007; Carlos Santiso, "Good Governance and Aid Effectiveness: The World Bank and Conditionality", The Georgetown Public Policy Review vol. 7 no. 1 (Fall 2001): 2.

[32] Daniel Kaufmann, Aart Kraay, Pablo Zoido–Lobatón, *Aggregating Governance Indicators*, The World Bank Policy Research Working Paper 2195 (October 1999), pp. 1–2, 5–7.

approach, however, has been criticized as being fragile and it has frequently been described as being either inadequate or unacceptable.[33] It is, therefore, difficult to establish what elements do contribute to good governance, according to the World Bank approach. Nevertheless, based on the aforesaid document, it is possible to conclude that the World Bank specifies four elements of good governance, these being: *accountability, effectiveness, rule of law* and *transparency*. Responding to the accusation of their concept lacking sufficient detail, they have responded by indicating six additional dimensions of good governance indicators, which are detailed below:

1. *Voice and accountability*—measuring political, civil and human rights
2. *Political Iistability and violence*—measuring the likelihood of violent threats to, or changes in, government, including terrorism
3. *Government effectiveness*—measuring the competence of the bureaucracy and the quality of public service delivery
4. *Regulatory burden*—measuring the incidence of market-unfriendly policies
5. *Rule of law*—measuring the quality of contract enforcement, the police, and the courts, as well as the likelihood of crime and violence
6. *Control of corruption*—measuring the exercise of public power for private gain, including both petty and grand corruption and state capture."[34]

Building upon the approach of the World Bank, the Asian Development Bank (ADB) intends to formulate an analytical framework for addressing governance issues and to draw a distinction between the elements of good governance.[35] Accordingly, the ADB, therefore, identifies four elements of good governance namely: *accountability, participation, predictability* and *transparency*.[36] Today, this definition is still largely quoted by the ADB.[37]

It should also be noted that the International Monetary Fund (IMF) has defined good governance as "the transparency of government accounts, the effectiveness of public resource management, and the stability and transparency of the economic and regulatory environment for private sector activity."[38] This suggests that the IMF emphasizes four fundamental elements of good governance, these being: *accountability, transparency, effectiveness* and *the rule of law*.

---

[33] Carlos Santiso, supra note 31, pp. 5–6.

[34] Daniel Kaufmann, Aart Kraay and Massimo Mastruzzi, "Measuring Governance Using Cross-Country Perceptions Data", The World Bank (August 2005), pp. 5, available from: http://siteresources.worldbank.org/INTWBIGOVANTCOR/Resources/GovMatters_IV_main.pdf   last visited date: 23/12/2007.

[35] Asian Development Bank, *Governance: Sound Development Management*, ISBN: 971-561-262-8 (August 1995), pp. 8.

[36] *Ibid.*

[37] See ADB website: http://www.adb.org/Governance/gov_elements.asp last visited date: 19/06/2008.

[38] The International Monetary Fund, *Good Governance—The IMF's Role*, ISBN: 1-55775-690-2 (August 1997), Introduction.

The World Bank and the IMF have been acting independently but each has been working closely with the United Nations Development Programme (UNDP). UNDP has been at the forefront of the development of international consensus that good governance and sustainable human development are integrally connected.[39] Originally, UNDP stated that much had been written regarding the features of efficient government, successful business and effective civil society organisations, however, the characteristics of good governance remain elusive.[40] They then went further and explicitly explore nine elements of good governance which appear to be a good definition and these being: *participation, rule of law, transparency, responsiveness, consensus orientation, equity, effectiveness and efficiency, accountability* and *strategic vision*. This definition sets the basis for other definitions. For example, in addressing gender sensitivity, UNDP emphasised seven elements of good governance namely: *participation, representation, accountability, transparency, responsiveness, efficiency* and *equity*.[41] From this, it is evident that, even within the same institution, slightly different elements have been chosen to tackle different governance issues. Apart from the above mentioned bodies, others have also been active in deliberating about what constitute good governance.

In the report of the United Nations Committee for Development Planning (UNCDP), some attributes of good governance have been identified. They recognize that, if a decision taken is *consensus based*, this might reduce the potential conflicts among the different interest groups, possibly resulting in peaceful regime change and institutional renewal. *Accountability* is identified as a crucial element in ensuring that political leaders and the bureaucracy delivers policy that meets the needs of the public. An open decision making system is also encouraged by the UNCDP, as this would give the public an opportunity to *participate* and express their views. They also opine that opening the decision making processes will also move towards the aim of *transparency* and an impartial system of law is required to be an *equitable and inclusive* in settling disputes, in order to protect personal security. *The rule of law* is considered as being essential, as this would provide an open, stable and predictable legal and political decision making system. Finally, political devolution and administrative *decentralization* is felt to be essential for the purpose of achieving more *coherent* governance. Furthermore, if the UNCDP's ultimate goal is to place the public interest as being of paramount concern, they will then require the public authority to *respond* to the needs of the public. To summarize, nine elements of good governance can be concluded, these being: *consensus based decision making, coherence, accountability, participation, transparency, rule of law, equity and inclusiveness, decentralization* as well as *responsiveness*. [42]

---

[39] UNDP, *Governance for Sustainable Human Development* (1997), available from: http://mirror.undp.org/magnet/policy/ last visited date: 23/12/2007.

[40] *Ibid.*

[41] UNDP, *Measuring Democratic Governance—A Framework for Selecting Pro-Poor and Gender Sensitive Indicators*, May 2006, p. 7.

[42] The Report of the United Nations Committee for Development Planning, *Poverty Alleviation and Sustainable Development: Goals in Conflict?* (1992), pp. 62–63.

In 2000, the commission of the European communities published a white paper on governance which proposed that "'governance' will be taken to encompass rules, processes and behavior that affect the way in which powers are exercised at European level, particularly as regards accountability, clarity, transparency, coherence, efficiency and effectiveness".[43] This white paper proposed six elements of good governance, namely, *accountability, clarity, transparency, coherence, efficiency and effectiveness.*[44] Having discussed a variety of views of various bodies regarding the concept of good governance, the paper will now turn to the academic attitude as to what might be considered to constitute good governance.

Academia views the issue of good governance from various perspectives, either by applying the concept to the specific areas[45] such as, electricity management[46] and corporate governance,[47] or discussing such specific shortcomings such as corruption[48] or secrecy.[49] The elements of good governance adopted in the academic work also vary considerably. For example, Melissa Dasgupta used "*public*

---

[43] Commission of the European Communities, *White Paper on European Governance—Enhancing Democracy in the European Union*, Brussels, 11th October 2002, SEC (2000) 1547/7 final, pp. 4; see also European Union, *Governance in the European Union: A White Paper—Enhancing Democracy in the European Union*, COM (2001) 428 final; European Commission, Report from the Commission on European Governance European Commission, 2003; see also European Commission, *Toward a Reinforced Culture of Consultation and Dialogue—General Principles and Minimum Standards for Consultation of Interested Parties*, COM (2002) 704 final.

[44] Commission of the European Communities, *White Paper on European Governance—Enhancing Democracy in the European Union* (October 2000), Brussels, 11th October 2002, SEC (2000) 1547/7 final, p. 4.

[45] Dr. Jonathan Benjamin-Alvarado, "Sustainability, Energy Policy and Future Good Governance in Cuba", Florida Coastal Law Journal (2001–2002): 430–431; Erik Luna, "Cuban Criminal Justice and the Ideal of Good Governance", Transnational Law & Contemporary Problems 14 (2004–2005): 529–755; Ngaire Woods, "Good Governance in International Organizations", Global Governance, no. 5 (1999): 39–61; C. M. G. Himsworth and C. M. O. 'Neill, *Scotland's Constitution: Law and Practice* (UK: LexisNexis, 2003), pp. 34–51.

[46] Francis N. Botchway, "The Role of the States in the Context of Good Governance and Electricity Management: Comparative Antecedents and Current Trends", University of Pennsylvania Journal of International Economic, vol. 21:4 (2000): 782–833.

[47] David Ablen. "The Principles of Good Corporate Governance and the Best Practice Recommendation's of the ASX Corporate Governance Council", Sydney Law Review (2003): 555–567; Sandeep Parekh, "Prevention of Insider Trading and Corporate Good Governance in India", International Business Lawyer, no. 32 (2004): 132–142; Bradley A. Helms and Richard H. Koppes, "Statistical Alchemy: How Methodological Shortcomings in the Inquiries into the Financial Impact of Corporate Governance Reform Prevent the Wall Streets of the World from Reaching a Consensus about the Value of Good Corporate Governance", Business Law International (2000): 204–222.

[48] Fidel V. Ramos, "Good Governance Against Corruption", The Fletcher Forum of World Affairs (2001): 9–19; C. Raj Kumar, "Corruption, Human Rights, and Development: Sovereignty and State Capacity to Promote Good Governance", American Society of International Law Proceeding, no. 99 (2005): 416–420.

[49] Saladin Al-Juef, "Good Governance and Transparency: Their Impact on Development", Transnational Law & Contemporary Problems, no. 9 (1999): 193–217.

access to environmental information"; "public participation in the legislative process" and *"judicial review"*, to assess good governance.[50] In addition, Francis N. Botchway indicated that good governance comprised four elements: *"democracy", "rule of law", "discretion"* and *"decentralization"*.[51]

Thomas M. Franck adopted four criteria to define or identify legitimacy in the international system, namely, *"determinacy", "symbolic validation", "coherence"* and *"adherence"*.[52] These criteria appear to be quite different comparing with the elements of good governance used by other writers. A closer looked at the criteria used by Franck, there was a clear linkage with good governance. Franck further explains that *determinacy* is what makes the rule's message clear or transparent,[53] thus confirming the importance of publishing laws in a comprehensible manner. The decision making processes also need to be transparent. *Symbolic validation* affirms and reinforces the importance of real equality.[54] It is a governance tool used by the public authorities to elicit compliance with a command. For example, a person's national identification is a legal signal symbolically reinforcing the citizen's relationship to the state and a relationship of rights and duties.[55] A citizen of a state should be legally equal, irrespective of any sexual distinction, religion, race, social class or political persuasion. This equality of participation is itself the symbolic representation of a confluence between different interest groups. In order to achieve the aforesaid objective,' the decision making processes needs to taking into account all the relevant different interests and voices. *Coherence* requires a rule to be applied consistently in every similar or applicable instance.[56] Following on from the above, a public authority has to follow an open rule of law in making decisions and the rules should be equally and inclusively applied as appropriate. *Adherence* addresses the importance of a rule to be made within the procedural and institutional framework.[57] The ethos of following the rule of law is once again emphasized and the law needs to be equally and inclusively applied by every public authority.

Peter Rogers and Alan W Hall state six necessary conditions for good water governance, and these being: *inclusiveness, accountability, participation,*

---

[50] Melissa Dasgupta, "The Access Initiative: Promoting Sustainable Development Through Good Governance", Sustainable Development Law & Policy (2004): 31–37.

[51] Francis N. Botchway (2001), supra note 16: 180–208.

[52] Thomas M. Franck, 'Legitimacy in the International System', The American Journal of International Law, Vol. 82 1988, p 712–759; Thomas M. Franck, 'The Power of Legitimacy and the Legitimacy of Power: International Law in an Ago of Power Disequilibrium', Vol. 100 2006, p 93.

[53] Thomas M. Franck, "The Power of Legitimacy and the Legitimacy of Power: International Law in an Ago of Power Disequilibrium", *Ibid*: 94.

[54] Thomas M. Franck, 'Legitimacy in the International System', The American Journal of International Law, p 725–735.

[55] *Ibid*, p 725.

[56] *Ibid*: 741.

[57] *Ibid*: 751–759.

transparency, predictability and *responsiveness*,[58] whereas, Patrick Molutsi establishes a set of mediating values for democratic governance: *participation, authorization, representation, accountability, transparency, responsiveness* and *solidarity*.[59] Marie Besançon states that the outcomes of good governance should create greater *transparency* within a *legitimate* system and capacity for measurement and *accountability*.[60]

The above debate suggests that different institutions or individuals understand and address the concept of good governance in different ways. It may, therefore, be inappropriate to try to promote universal and fixed elements of good governance. Some of the authors or institutions adopted 'bad governance indicators', in order to assess the overall economic performance of a country in aspects such as *graft*,[61] *political instability and violence*,[62] so as. These 'bad governance indicators' are helpful in identifying the specifics of good governance. For example, a *transparent* government decision making process would be expected to reduce the degree of *graft*. *Political instability* can be overcome by good communication and by considering fairly, the different interests in the society which draws attention to the importance of *public participation, equitable and inclusive* as well as *coherent*, when considering good governance. A sound system of the *rule of law* is expected to institute a dispute settlement body which will in turn eliminate *violence* behaviours and other misdemeanours. Notwithstanding the above, this research will not consider 'bad governance indicators' further as the relevant issues will be detailed while discussing the elements of good governance.

This chapter lists the following elements of 'good governance' which have recurred frequently both in the literature and in political and in practitioner debates on the subject:

1. *Accountability;*
2. *Effectiveness and efficiency;*
3. *Rule of law;*
4. *Transparency;*
5. *Public access to environmental information;*
6. *Regulatory burden;*
7. *Control of corruption;*
8. *Participation;*
9. *Predictability;*

---

[58] Peter Rogers and Alan W Hall, *Effective Water Governance*, Global Water Partnership Technical Committee, The Background Papers No.7 (February 2003), p. 9.

[59] Patrick Molutsi, supra note 31: 5.

[60] Marie Besançon, *Good Governance Rankings: The Art of Measurement*, World Peace Foundation Report, Number 36 (2003), p. 6–10.

[61] Daniel Kaufmann, Aart Kraay, Pablo Zoido-Lobatón, *Aggregating Governance Indicators*, The World Bank Policy Research Working Paper 2195, October 1999.

[62] Daniel Kaufmann, Aart Kraay and Massimo Mastruzzi, 'Measuring Governance Using Cross-Country Perceptions Data', The World Bank, August 2005.

10. *Responsiveness;*
11. *Consensus orientation;*
12. *Equity and inclusiveness;*
13. *Strategic vision;*
14. *Representation;*
15. *Coherency;*
16. *Clarity;*
17. *Democracy;*
18. *Discretion;*
19. *Decentralization;*
20. *Determinacy;*
21. *Symbolic validation;*
22. *Adherence;*
23. *Authorization;*
24. *Solidarity.*

Based on the aforementioned literature[63] and by applying a mathematical approach, there are 11 authors or institutions which emphasize the importance of *accountability* and *transparency* in good governance. Eight authors and institutions mentioned the ethos of *rule of law* and seven authors and institutions have emphasized the need to open the decision making processes to public *participation*. Six authors and institutions mentioned the element of *effectiveness and efficiency* as a part of good governance while four noted that it is necessary for the public authority to *respond* to the needs of the public/environment. Three have emphasized the importance of *equity and inclusiveness*, and *coherence* to the concept of good governance, whereas, two have adopted the elements of *predictability, consensus based decision making, representation, decentralization* in order to assess the performance of good governance.

Only one author or institution has mentioned the elements of *public access to environmental information*; *regulatory burden*; *control of corruption*; *strategic vision*; *clarity*; *democracy*; *discretion*; *determinacy*; *symbolic validation*; *adherence*; *authorization* as well as *solidarity*. Nonetheless, some of these elements only being mentioned once are broadly similar to those defined as being more frequently utilized. For example, *public access to environmental information* can be considered as being a part of *transparency*; a sound concept of the *rule of law* covers *regulatory burden, clarity, discretion* as well as *adherence*; by *transparency* and ensuring that the *rule of law* is followed will achieve *determinacy* and will also enhance the *control of corruption*; *symbolic validation* is similar to the element of *equitable and inclusive* and finally *democracy* can be achieved by opening the decision making processes to public participation.

---

[63] This chapter does not intend to provide an exhaustive literature overview in relation to good governance. What has been presented here is largely based on the available literature and documents at the time of writing.

In order to employ the aforementioned elements of good governance as a workable analytical framework, it is necessary to narrow down the scope of discussion and eliminate part of the elements mentioned in the literature. Given the fact that these elements which have been mentioned only once by various authors or institutions are very similar to these which have been mentioned more than once. This chapter, therefore, will not consider these elements which have been mentioned only once separately. The aim is to arrive at a more used list in terms of the element of good governance, the latter comprising, *accountability, transparency, rule of law, participation, effectiveness and efficiency, responsiveness, equity and inclusiveness, coherence, predictability, consensus based decision making, representation* and *decentralization.*

A closer scrutiny of these elements drawn from various authors and institutions would indicate that there appears to be little convergence and thus, further discussion regarding the selection of the elements of good governance is, therefore, necessary. Although a number of authors and institutions suggest that *effectiveness and efficiency* is a crucial element of good governance, this would seem more likely to be the outcome of the overall performance of good governance, rather than components thereof. This is because an effective and efficient decision making needs to meet the needs of society while making the best use of the resources at their disposal. This may be achieved only when there is a 'good' governance mechanism in place. It is also very difficult to establish how effective and efficient a decision making system is. Based on the above, this chapter will not employ *effectiveness and efficiency* as an integral part of good governance. The element of *predictability* is covered by the sound concept of *rule of law*, which should lead to an open and stable legal system. The net result should provide a predictable decision making. Apart from the aforementioned, a transparent decision making system will also lead to greater predictability. The issues in relation to *predictability* will, therefore, be considered under the heading of *rule of law* and *transparency.* The element of *participation* should include the ethos of *representation,* indicating that the public should be given the opportunity to represent themselves in the decision making processes. The element of *representation* will not be considered separately. The elements of *decentralization* will be considered under the heading of *coherence,* as it relates to the institutional responsibility and the procedure to achieve a better co-ordination.

As a result of the above debate, the elements of good governance can be further reduced for further consideration, these elements being as follows, the *rule of law, participatory, transparency, consensus based decision making, accountability, equitable and inclusive, responsive* and *coherent.* Figure 1.1 indicates a fundamental fact that the characteristics or indicators of good governance are mutually reinforcing and cannot stand alone. The significance of these principles and processes of public governance can be expected to vary between contexts and over time.[64]

---

[64] Tony Bovaird and Elke Löffler, "Evaluating the Quality of Public Governance: Indicators, Models and Methodologies", International Review of Administrative Science (2003): 322.

Having considered the various aspects of good governance, the paper will now examine the various different methods of measurement which may be employed in relation to the concept. There has been a significant amount of work aimed at measuring governance by means of an indicator based approach.[65] Some of the governance indicators refer to non-commercial indices such as Freedom House's indices, Transparency International's corruption perception index, the Polity IV and Polyarchy datasets.[66] These indicators are increasingly used by international investors to gauge business opportunities and foresee major crises in monetary terms. Freedom House's indices and Transparency International's Corruption Perception Index are embedded in the ethical values that they aim to promote. The aforementioned indicators are, in turn, full of measurement and comparability problems, therefore, making them unsuitable for cross-sector assessment.[67] Although these indicators do not directly point out the elements of good governance, a closer scrutiny reveals that they actually stand upon the elements of 'the rule of law', 'transparency' as well as 'accountability'.[68] The Polity IV and Polyarchy datasets are said to be more accurate and refined measures of governance and democracy.[69] Once again, they do not directly point out the elements of good governance. A closer reading of the indices suggests that they are helpful to establish the elements of 'participatory', 'transparency' 'accountability' and 'the rule of law' in good governance.[70]

---

[65] Cris Whitehouse, "The Ants and the Cockroach—A Challenge to the Use of Indicators", available from: http://www.mande.co.uk/docs/Indicators%20-%20The%20Ants%20and%20the%20Cockroach.pdf last visited date: 18/03/2008.

[66] Anja Linder and Carlos Santiso, "Assessing the Predictive Power of Country Risk Ratings and Governance Indicators", SAIS Working Paper Series, WP/02/02, p. 4–5, available from: http://www.sais-jhu.edu/workingpapers/WP-02-02b.pdf last visited date: 23/12/2007; The World Bank, "Toward more Operationally Relevant Indicators of Governance", Number 49, December 2000, p. 1–4; P. Norris, "Governance Indicators", in *Driving Democracy: Do Power Sharing Institutions Work* (Cambridge University Press, London 2006), available from: http://ksghome.harvard.edu/~pnorris/Acrobat/Driving%20Democracy/Chapter%203.pdf. Last visited date: 23/12/2007; The Hungarian Gallup Institute, "Basic Methodological Aspects of Corruption Measurement: Lessons Learned From the Literature and the Pilot Study" (December 1999), available from: http://www.unodc.org/pdf/crime/corruption_hungary_rapid_assess.pdf last visited date: 23/12/2007; Gerardo L. Munck, "Measures of Democracy, Governance and Rule of Law: An Overview of Cross-National Data Sets", paper prepared for world bank workshop, available from: http://lnweb18.worldbank.org/ESSD/sdvext.nsf/68ByDocName/Munck/$FILE/Munck+Paper.pdf last visited date: 23/12/2007; Julius Court, Goran Hyden and Ken Mease, "Assessing Governance: Methodological Challenges", World Governance Survey Discussion Paper 2 (August 2002), available from: http://www.odi.org.uk/wga_governance/Abstracts/Publ_Abs_02.html last visited date: 23/12/2007.

[67] Anja Linder and Carlos Santiso,supra note 66: 4.

[68] The World Bank, *Toward more Operationally relevant Indicators of Governance*, No. 49 (December 2000), available from: http://www-wds.worldbank.org/external/default/WDSContentServer/WDSP/IB/2001/03/01/000094946_01021005390645/Rendered/PDF/multi0page.pdf last visited date: 19/06/2008.

[69] Anja Linder and Carlos Santiso, supra note 66: 5.

[70] *Ibid.*

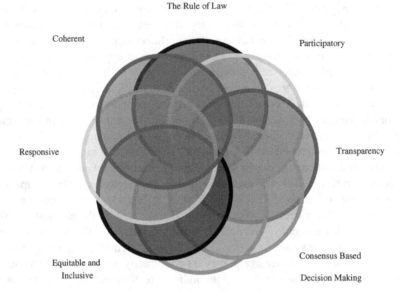

Fig. 1 Characteristics of Good Governance, (**source** Designed by Yen-Chiang Chang)

Different institutions or writers have adopted different indicators, in order to tackle different issues. The Organisation for Economic Co-operation and Development (OECD) emphasizes the need to combine the use of "input", "output" and "intermediary outcome" indicators in conjunction for better determination of the efficiency and effectiveness of enforcement programs.[71] A study of the OECD documents would, however, seem to indicate that they still have doubts as to the necessary components of good governance. The use of the aforesaid indicators aims to identify the measurement gaps which in turn stimulate the development of more appropriate governance tools and systems for performance evaluation.[72] The net result is likely to lead to 'good governance'.

In order to employ 'good governance' as a legal analytical framework to assess State practice on ocean governance, further clarification in terms of the usage of the terms is necessary. An effective ocean governance system places further requirements on the process of decision-making and public policy formulation. It extends beyond the capacity of the public sector to the rules that create a legitimate

---

[71] OECD, "Summary of the INECE-OECD Workshop on Environmental Compliance and Enforcement Indicators: Measuring What Matters", available from: http://www.inece.org/indicators/workshop.html last visited date: 23/12/2007; OECD, "OECD Key Environmental Indicators" (2004), p. 7–8, available from: http://www.oecd.org/dataoecd/32/20/31558547.pdf last visited date: 23/12/2007.

[72] *Ibid.*

and coherent framework for the conduct of national marine policy. It implies governing public matters in a transparent, accountable, participatory, responsive, equitable and inclusive manner. It entails sufficient public participation in order to achieve a broad consensus in a society and an independent judiciary, institutional balances through a horizontal and vertical separation of powers and having effective supervisory agencies. The next part of this chapter will provide examples from practice for each element of good governance which depicted earlier in this part of discussion. The aim is to prove that these elements of good governance are supported by State practice.

1.3 State Practice in Relation to Each Element of Good Governance

The Rule of Law. Different people understand the meaning of the rule of law in different ways, for example, the UNDP interprets the concept of rule of law as "Legal frameworks should be fair and enforced impartially,..."[73] The ADB goes further to state that "The rule of law encompasses both well-defined rights and duties, as well as mechanisms for enforcing them, and settling disputes in an impartial manner."[74] It must be said, however, that every person, irrespective of rank and status in society, should be subject to the law.[75] As stated by the 1992 Convention on Biological Diversity, legislative, administrative or policy measures shall be taken to ensure the following of international law[76] in "a fair and equitable way."[77]

---

[73] UNDP, *Governance for Sustainable Human Development* (1997), available from: http://mirror.undp.org/magnet/policy/ last visited date: 02/01/2008. Similar opinions please see Daniel Kaufmann, Aart Kraay and Massimo Mastruzzi, "Measuring Governance Using Cross-Country Perceptions Data", The World Bank (August 2005), p. 5, available from: http://siteresources.worldbank.org/INTWBIGOVANTCOR/Resources/GovMatters_IV_main.pdf last visited date: 23/12/2007; The Report of the United Nations Committee for Development Planning, *Poverty Alleviation and Sustainable Development: Goals in Conflict?* (1992), p. 62–63; Commission of the European Communities, *White Paper on European Governance—Enhancing Democracy in the European Union* (October 2000), p. 4; Melissa Dasgupta, supra note 50: 31–37; Francis N. Botchway (2001), supra note 16,:197–198; Thomas M. Franck, "The Power of Legitimacy and the Legitimacy of Power: International Law in an Ago of Power Disequilibrium", supra note 52: 94.

[74] Asian Development Bank, *Governance: Sound Development Management*, ISBN: 971-561-262-8 (August 1995), pp. 8.

[75] Joseph Raz, "The Rule of Law and Its Virtue", Law Quarterly Review no. 93 (1997): 1; Neil Parpworth, *Constitutional and Administrative Law* (Edinburgh: LexisNexis UK, 2004), pp. 34; Hilaire Barnett, *Constitutional & Administrative Law* 4th ed. (London: Cavendish, 2002), pp. 69; Francis N. Botchway (2001), supra note 16: 197.

[76] Article 16 (3) of the 1992 Convention on Biological Diversity.

[77] Article 15 (7) of the 1992 Convention on Biological Diversity.

Following the rule of law is a yardstick by which to measure "both the extent to which government acts under the law and the extent to which individual rights are recognized and protected by law."[78] The 1992 Rio Declaration, for example, states that "States shall enact effective environmental legislation. Environmental standards, management objectives and priorities should reflect the environmental and development context to which they apply."[79] In this respect, all laws should be guided by open, stable, clear and general rules and an independent, easy access courts' review system should also be guaranteed.[80]

For law to be effective, it must be promulgated appropriately for example via the internet, newspapers or national television, particularly for those who are to be guided or affected by the law.[81] Where multiple languages are used in a country, it should be considered essential to publish the laws in all of the appropriate languages, for example, in Canada, law is published in both English and French;[82] in Wales, law should be published in both English and Welsh.[83]

In addition, both the legal system and individual laws must promote stability but this does not necessarily mean that the relevant laws remaining permanently unaltered. It is, however, unrealistic to expect the law to cover every conceivable circumstance or fact of life.[84] To sustain the long term stability of the legal system, it is necessary to have general principle legislation which allows the opportunity for legislative and executive authorities to promulgate more detailed subsidiary legislation.[85] This is evident from the 1982 United Nations Convention on the Law of the Sea (UNCLOS), which allows states to decide "the best practicable means at their disposal and in accordance with their capabilities"[86] to prevent and control marine pollution.

Subsidiary legislation may be more suitable and convenient for the regulation of ocean governance activities, as it can be more easily amended to reflect current and diverse needs, without affecting the stability of the principal legislation.[87] Thus, subsidiary legislation allows for laws to be passed when new situations arise, thereby allowing the new law to reflect the current situation.[88] In the light of the aforesaid, the element of 'rule of law' inevitably relates to the element of

---

[78] Hilaire Barnett (2004), supra note 75, pp. 69.

[79] Principle 11 of the 1992 Rio Declaration.

[80] Neil Parpworth (2004), supra note 75, pp. 34; Hilaire Barnett (2004), supra note 75, pp. 69.

[81] Francis N. Botchway (2001), supra note 16: 197.

[82] Section 18, Schedule B, Constitution Act 1982, enacted as Schedule B to the Canada Act 1982 (U.K.), c.11, came into force on 17th April 1982.

[83] Section 47, Government of Wales Act 1998, c.38, 1998 Chapter 38.

[84] Francis N. Botchway (2001), supra note 16: 200–201.

[85] Francis N. Botchway (2001), supra note 16: 197–198.

[86] Article 194.1 of UNCLOS.

[87] Urias Forbes, "Subsidiary Law-Making Process: Antigua, Dominica & St. Kitts", International and Comparative Law Quarterly, vol. 18 (1969): 533–534.

[88] Francis N. Botchway (2001), supra note 16: 201.

'responsive' which requires the public authorities to respond to the needs of the public and the environment. There is, however, a limit to the usage of subsidiary legislation, in that, the subsidiary legislation must not derogate from or be inconsistent with, the principal legislation.[89]

The Oceans Act 1996[90] in Canada, is an example which can offer a legal basis for the aforesaid concept. In exercising the powers and performing the duties and functions mentioned in the Oceans Act 1996, the Minister of Fisheries and Oceans shall "develop and implement policies and programs with respect to matters assigned by law to the Minister."[91] The Minister of Fisheries and Oceans shall also "coordinate with other ministers, boards and agencies of the Government of Canada the implementation of policies and programs of the government with respect to all activities or measures in or affecting coastal waters and marine waters."[92] This provision suggests that the Oceans Act 1996 assigned all activities or measures in or affecting coastal and marine waters to the Minister of Fisheries and Oceans. The Minister should, within his/her authority develop and implement appropriate policies and programs, including enacting subsidiary legislation. The said conduct should be within the scope of the Oceans Act 1996 and compatible with it.

It is therefore clear that, for the purposes of examining the element of the rule of law, the following issues must be considered. Firstly, must all of the laws have to be published in an appropriate manner? Secondly, can the law be equally and fairly administered and effectively enforced? Thirdly, are decision makers and administrators bound to follow the rule of law when making decisions?

Participatory. There is a growing concern regarding the need for public participation in environmental law, which can be seen in relation to both international law and domestic law.[93] An example of this is illustrated in the 1992 Rio Declaration, which states that "States and people shall cooperate in good faith and in a spirit of partnership in the fulfillment of the principles embodied in this Declaration..."[94] The requirement is that, government must ensure an open decision-

---

[89] Article 11 of the Act of Central Legislations Standards, Taiwan, announced by President in 31st August 1970, revised on 19th May 2004, President order number: 09300094181.

[90] The Oceans Act 1996, c. 31, Assented to 18th December 1996.

[91] Section 32 (a) of the Oceans Act 1996.

[92] Section 32 (b) of the Oceans Act 1996.

[93] For international law, the example is the Convention on Access to Information, Public Participation in Decision-Making and Access to Justice in Environmental Matters (Aarhus) 25th June 1998, which came into force on 30th October 2001, 38 ILM 517 (1999). For domestic law, evidence can be seen in the Pollution Prevention and Control (Public Participation ect.) (Scotland) Regulations 2005, SSI 2005 No. 520, 16th November 2005; the Pollution Prevention and Control (Designation of Public Participation Directive) (Scotland) Order 2005, SSI 2005 No. 461, 5th October 2005.

[94] Principle 27 of the 1992 Rio Declaration.

making system in which public interests are fully respected.[95] As the UNDP emphasizes that "All men and women should have a voice in decision-making, either directly or through legitimate intermediate institutions that represent their interests."[96]

The Convention on Access of Information, Public Participation in Decision-Making and Access to Justice in Environmental Matters,[97] (1998 Aarhus Convention) defines public participation as: giving members of the public "the opportunity to express their concerns and enable[ing] public authorities to take due account of such concerns."[98] This process can be understood as involving interactive activities through which individuals and private groups not holding government authority seek to influence policy, regulatory,[99] together with those policy-influencing interactions of government officials that go well beyond the direct use of their authority.[100] This legal obligation requires the law to provide the general public with the rights to obtain information which will enable them to question, challenge or otherwise influence decision-making.[101] As Chapter 17 of Agenda 21 states, where ever possible, "for concerned individuals, groups and organizations to relevant information and opportunities for consultation and participation in planning and decision-making at appropriate levels is necessary."[102] Following on from the above, the element of 'participatory' relates to the element of 'transparency', which requires the public authorities to release certain obtained information.

---

[95] Asian Development Bank, *Governance: Sound Development Management*, ISBN: 971-561-262-8 (August 1995), pp. 8; UNDP, *Measuring Democratic Governance—A Framework for Selecting Pro-Poor and Gender Sensitive Indicators* (May 2006), pp. 7; The Report of the United Nations Committee for Development Planning, *Poverty Alleviation and Sustainable Development: Goals in Conflict?* (1992), pp. 62–63; Melissa Dasgupta, supra note 50: 31–37; Peter Rogers and Alan W Hall, supra note 58: 9; Patrick Molutsi, supra note 31: 5.

[96] UNDP, *Governance for Sustainable Human Development* (1997), available from: http://mirror.undp.org/magnet/policy/ last visited date: 02/01/2008.

[97] The Convention on Access to Information, Public Participation in Decision-Making and Access to Justice in Environmental Matters (Aarhus) 25th June 1998, came into force on 30th October 2001, 38 ILM 517 (1999).

[98] Preamble of the 1998 Aarhus Convention.

[99] Article 3.1 of the 1998 Aarhus Convention; see also Kirsty Blackstock, Elizabeth Kirk, Yen-Chiang Chang and Grant Davidson, *Public Participation and Consultation in SEPA Regulatory Regimes*, Published by Scottish Environment Protection Agency, March 2006, available from: http://www.sepa.org.uk/publications/for_sepa.htm last visited date: 16/06/08.

[100] Charles E. Lindblom and Edward J. Woodhouse, *The Policy-Making Process* 3rd ed. (New Jersey : Prentice Hall, Englewood Cliffs, 1993) , p 75.

[101] Three core objectives were stated by the 1998 Aarhus Convention: 1. promoting public access to environmental information; 2. promoting increased public participation in environmental decision-making by establishing a fair and transparent framework for decisions; 3. promoting access to justice on environmental matters by giving the public the right to question the decisions made by the court or public authorities.

[102] Paragraph 17.5 (f) of the Agenda 21.

Public participation, on the one hand, is the requirement to offer the opportunity to the general public to express their opinions, which could constrain the public authorities in the decision-making process. On the other hand, due to the diversity of public perspectives and to their sources, it is important to be 'equitable and inclusive' when making decisions.[103] This conduct may, in turn, lead to a 'consensus based decision-making' process. The element of 'participatory' is, therefore, connected to other elements. There are various ways to achieve public involvement, such as consultation, public survey or public inquiries.

An open style of government may allow parliament, interest groups not holding government authority and individuals to participate in decision-making. A government which is more open is likely to be better informed than a government that is restrictive. This openness should consequentially lead to a better quality of decisions. This is because participation in the democratic decision making process should not end with an election vote but should "continue to allow citizens the opportunity to contribute to the system of government decision-making."[104] In support of this, the ADS states "The principle of participation derives from an acceptance that people are at the heart of development."[105]

Examples of this aspect can be found in State practice. The US Oceans Act 2000 insists that the commission on Ocean Policy opens all meetings to the public and therefore, interested persons shall be permitted to appear at open meetings and present oral or written statements on the subject matter of the meeting.[106] In Canada, the Minister of Fisheries and Oceans, for the purposes of the implementation of integrated governance plans may, at his/her discretion, consult "another person or body or with another minister, board or agency of the Government of Canada, and taking into consideration the views of ...other persons and bodies, including those bodies established under land claims agreements."[107]

In order to set up criteria for a further examination of public participation, this research firstly suggests focusing on whether legislation imposes a legal obligation requiring public involvement in the decision-making process. Secondly, the research suggests considering the form in which the public participation is organized and whether this is satisfactory in meeting the criteria for public participation.

---

[103] Jenny Steele, "Participation and Deliberation in Environmental Law: Exploring a Problem-Solving Approach", Oxford Journal of Legal Studies, vol. 21, no. 3 (2001): 416, 441.

[104] John F. McEldowney, Public Law (London: Sweet & Maxwell, 2002), pp. 574.

[105] Asian Development Bank, Governance: Sound Development Management, ISBN: 971-561-262-8 (August 1995), pp. 9.

[106] Section 3 (e) (1) of the Oceans Act 2000, S. 2327, PL 106-256, 25th July 2000.

[107] Section 32 (c) (d) of the Oceans Act 1996.

Transparency: Transparent decision-making requires an open process that allows the general public's monitoring and participation,[108] which means that any decision taken must follow the rule of law.[109] The general public and public authorities need to know what kinds of decisions can be opened to the public and how the decision-making process should progress. Therefore, law must be able to include all the aforesaid information. Following the rule of law and public participation have already been discussed but it is further stressed that the decision-making process must be open to the public and information should be freely available and accessible to those who will be affected by such decisions and their enforcement.[110] Individuals or interest groups must have the right to know why decisions are made, which requires the public authorities to state clear objectives

---

[108] See Carlos Santiso, supra note 31: 5–6; The International Monetary Fund, *Good Governance—The IMF's Role*, ISBN: 1-55775-690-2 (August 1997), Executive Summary; Asian Development Bank, *Governance: Sound Development Management*, ISBN: 971-561-262-8 (August 1995), pp. 8; UNDP, *Governance for Sustainable Human Development* (1997), available from: http://mirror.undp.org/magnet/policy/ last visited date: 23/12/2007; The Report of the United Nations Committee for Development Planning, *Poverty Alleviation and Sustainable Development: Goals in Conflict?* (1992), pp. 62–63; Commission of the European Communities, *White Paper on European Governance—Enhancing Democracy in the European Union*, Brussels, 11th October 2002, SEC (2000) 1547/7 final; European Union, *Governance in the European Union: A White Paper—Enhancing Democracy in the European Union*, COM (2001) 428 final; European Commission, Report from the Commission on European Governance (European Commission, 2003); see also European Commission, *Toward a Reinforced Culture of Consultation and Dialogue—General Principles and Minimum Standards for Consultation of Interested Parties*, COM (2002) 704 final; Melissa Dasgupta, supra note 50:31–37; Thomas M. Franck, 'The Power of Legitimacy and the Legitimacy of Power: International Law in an Ago of Power Disequilibrium', supra note 52: 94; Peter Rogers and Alan W Hall, supra note 58: 9; Patrick Molutsi, supra note 31: 5.

[109] The United Nations Economic and Social Commission for Asia and the Pacific, Human Settlements, "What is Good Governance?", available from: http://www.unescap.org/huset/gg/governance.htm last visited date: 05/06/2006. UNDP, *Governance for Sustainable Human Development* (1997), available from: http://mirror.undp.org/magnet/policy/ last visited date: 02/01/2008. Similar opinions please see Daniel Kaufmann, Aart Kraay and Massimo Mastruzzi, "Measuring Governance Using Cross-Country Perceptions Data", The World Bank (August 2005), pp. 5, available from: http://siteresources.worldbank.org/INTWBIGOVANTCOR/Resources/GovMatters_IV_main.pdf last visited date: 23/12/2007; The Report of the United Nations Committee for Development Planning, *Poverty Alleviation and Sustainable Development: Goals in Conflict?* (1992), pp. 62–63; Commission of the European Communities, *White Paper on European Governance—Enhancing Democracy in the European Union* (October 2000), p. 4; Melissa Dasgupta, supra note 50: 31–37; Francis N. Botchway (2001), supra note 16: 197–198; Thomas M. Franck, "The Power of Legitimacy and the Legitimacy of Power: International Law in an Ago of Power Disequilibrium", supra note 52: 94.

[110] The United Nations Economic and Social Commission for Asia and the Pacific, Human Settlements, "What is Good Governance?", available from: http://www.unescap.org/huset/gg/governance.htm last visited date: 05/06/2006.

for their decisions and they should be explicit regarding the criteria, rationale and considerations on which their decisions are based.[111]

The right of access to environmental information is emphasized by the weight which international law place on greater disclosure. As stated in the 1992 Rio Declaration, "each individual shall have appropriate access to information concerning the environment that is held by public authorities,..."[112] Furthermore, Chapter 17 of Agenda 21[113] states that coastal states should commit themselves to integrated governance and sustainable development of coastal areas and the marine environment, under their national jurisdiction. It is thus necessary to "provide access, as far as possible, for concerned individuals, groups and organizations to relevant information..."[114] This general commitment to greater access to environmental information is specified in various international conventions. In addition to the aforesaid, the purpose of the 1998 Aarhus Convention was to ensure greater consistency and transparency of public access to environmental information and it has, for example, led to public authorities making minutes of meetings available to the public.[115] Further examples of disclosure include, the scope, cost-benefit and other economic analyses and assumptions used in environmental decision-making.[116]

Concerning the complexity and inherent secrecy of the public authorities' decision-making process, the general public may not know what decision has been made until the information has been published. In the US, however, the minutes and records of 'ALL' meetings and other documents that were made available to or prepared for the commission on Ocean Policy are available for public inspection.[117] Prior to delivery of the final report, the draft report has to be made available for review by the public and the governors of coastal states.[118] The aforesaid conduct may further strengthen transparent decision-making. For a further examination of transparent decision-making in practice, the following issues should be considered: firstly, does legislation impose an obligation on public authorities to release environmental information including information on decision

---

[111] The Independent Commission for Good Governance in Public Services, *The Good Governance Standard for Public Services*, OPM and CIPFA (2004), pp. 15.

[112] Principle 10 of the 1992 Rio Declaration.

[113] United Nations, *Report of World Summit on Sustainable Development*, Johannesburg, South Africa (26 August–4 September 2002), ISBN 92-1-104521-5, pp. 8.

[114] Paragraph 17.5 of the Chapter 17 of Agenda 21.

[115] Article 4, 5 of the 1998 Aarhus Convention.

[116] Article 2 (3) (c) of the 1998 Aarhus Convention.

[117] Section 3 (e) (2) of the Oceans Act 2000.

[118] Section 3 (g) (1) (2) of the Oceans Act 2000.

making?[119] Secondly, in what form is the environmental information presented? Thirdly, are requests for information met?

Consensus Based Decision Making: The UNDP emphasizes the importance of "consensus-building", however, it does not clearly indicate how to achieve consensus.[120] Consensus based decision-making needs to be supported by an open system which allows the general public to freely examine the decision-making activities and provides the opportunity to express their concern, thus enabling public authorities to take due account of such concern.[121] To achieve the said objective, the decision makers should consider two other aspects of good governance, namely, 'participatory' and 'transparent', which were discussed earlier in this chapter. The other aspect which needs to be considered is the fact that different interests in society need to be consulted in order to reach a broad consensus on what is in the best interests of the whole community.[122] This suggests that as many stakeholders as possible should be involved in the decision making process. In addition to the above, the element of 'equitable and inclusive' must also be considered.

Consensus based decision-making means taking decisions on the basis of common agreement of all those involved in the decision-making process. All participants are, therefore, working as a group toward a general agreement on all major issues, in order to reach collective decisions.[123] Even although seeking consensus is desirable in good and appropriate government, on many occasions, the government will have to make decisions, with which some parties are not

---

[119] See Article 2.3 of the 1998 Aarhus Convention: " 'Environmental information' means any information in written, visual, aural, electronic or any other material form on:(a) The state of elements of the environment, such as air and atmosphere, water, soil, land, landscape and natural sites, biological diversity and its components, including genetically modified organisms, and the interaction among these elements; (b) Factors, such as substances, energy, noise and radiation, and activities or measures, including administrative measures, environmental agreements, policies, legislation, plans and programmes, affecting or likely to affect the elements of the environment within the scope of subparagraph (a) above, and cost-benefit and other economic analyses and assumptions used in environmental decision-making; (c) The state of human health and safety, conditions of human life, cultural sites and built structures, inasmuch as they are or may be affected by the state of the elements of the environment or, through these elements, by the factors, activities or measures referred to in subparagraph (b) above."

[120] UNDP, *Governance for Sustainable Human Development* (1997), available from: http:// mirror.undp.org/magnet/policy/ last visited date: 03/01/2008.

[121] Preamble of the 1998 Aarhus Convention. See also the Report of the United Nations Committee for Development Planning, *Poverty Alleviation and Sustainable Development: Goals in Conflict?* (1992), pp. 62–63; UNDP, *Governance for Sustainable Human Development* (1997), available from: http://mirror.undp.org/magnet/policy/ last visited date: 23/12/2007.

[122] The United Nations Economic and Social Commission for Asia and the Pacific, Human Settlements, "What is Good Governance?", available from: http://www.unescap.org/huset/gg/ governance.htm last visited date: 05/06/2006.

[123] National Marine Sanctuaries, Monterey Bay, "Joint Management Plan Review Working Group—Consensus Based Decision Making", available from: www.sanctuaries.noaa.gov/joint-plan/mb_docs/mb_consensus.pdf last visited date: 22/03/07.

happy. The core concern of consensus based decision making is that no individual, official or group should be able to force their individual decisions or views on others, whether through majority voting or otherwise.[124] Reflecting on this role, attention should be focused on whether or not the current legislation provides the requirement to obtain consensus.

Accountable: Accountability is a fundamental element of good ocean governance. The UNDP states accountability as "Decision-makers in government, the private sector and civil society organizations are accountable to the public, as well as to institutional stakeholders."[125] From a broader point of view, not only public authorities but also private organizations must be accountable to the general public and to their stakeholders. This chapter is, however, only concerned with the public authorities' decision-making processes, thus, decisions made by private organizations are outwith the scope of this discussion. Within public authorities, who is accountable and to whom, depends on whether decisions are taken internally or externally. Internal decisions are conducted under a code of practice operating within public authorities and usually relate to internal affairs. A good example of this is the use of information for the purpose of monitoring, governing, inspecting or prohibiting vocational work.[126] Internal accountability may prove effective but it is often hidden from external scrutiny and to some extent, this may perpetuate secrecy, even accepting that the nature of internal review involves confidential and sensitive information.[127] The general public is primarily affected by external decisions and public authorities are ultimately accountable to those who will be affected by their decisions.[128]

The ADB observes that "Accountability is imperative to make public officials answerable for government behavior and responsive to the entity from which they derive their authority."[129] The aforesaid objective may be approached differently

[124] *Ibid*; see also Dr. John Robert Dew, "Consensus Based Decision Making", available from: http://bama.ua.edu/ ~ st497/ppt/consensusbaseddecision.ppt last visited date: 08/06/2007.

[125] UNDP, *Governance for Sustainable Human Development* (1997), available from: http://mirror.undp.org/magnet/policy/ last visited date: 02/01/2008.

[126] Article 5.4 of the Regulations of Administrative Information Openness, Taiwan, Executive Yuan order number 008048 in conjunction with Examination Yuan order number 00569, 21st of February, 2001.

[127] John F. McEldowney, *Public Law*, (Sweet & Maxwell's Textbook Series, 2002), pp. 278.

[128] The United Nations Economic and Social Commission for Asia and the Pacific, Human Settlements, "What is Good Governance?" available from: http://www.unescap.org/huset/gg/governance.htm last visited date: 05/06/2006.

[129] Asian Development Bank, *Governance: Sound Development Management*, ISBN: 971-561-262-8 (August 1995), pp. 8. See also Carlos Santiso, supra note 32: 5–6; Daniel Kaufmann, Aart Kraay and Massimo Mastruzzi, supra note 31: 5; The International Monetary Fund, *Good Governance—The IMF's Role,* supra note 38, Ececutive Summary; UNDP, *Governance for Sustainable Human Development*, supra note 39, pp. 3; The Report of the United Nations Committee for Development Planning, *Poverty Alleviation and Sustainable Development: Goals in Conflict?*, supra note 42, pp. 62–63; Commission of the European Communities, *White Paper on European Governance—Enhancing Democracy in the European Union* (October 2000), p 4; Peter Rogers and Alan W Hall, supra note 58: 9; Patrick Molutsi, supra note 31: 5; Marie Besançon, supra note 60: 6–10.

in different countries or political structures. From a constitutional aspect, the work of government is usually carried out by departments which are staffed by officials and led by ministers, although, some tasks may be delegated to non-departmental bodies. For instance, in Taiwan, the Overseas Fisheries Development Council of the Republic of China has contributed to fisheries negotiations.

The doctrine of ministerial responsibility means that "the Minister is ultimately responsible for all activity in his or her department and is answerable to Parliament and to the public for the actions of his or her department."[130] Establishing adequate institutional structures which keep government accountable to the public is a fundamental concern of integrated oceans and coastal management.[131] While this may be achieved by reducing administrative burdens and producing more efficient and effective regulation, the stated objective is more a medium to long-term goal.

The issue of accountability, therefore, needs to be considered in relation to an open, clear, stable legal system and transparency in the decision-making process. The elements of 'rule of law' and 'transparency' in good governance thus impinge upon the issue of accountability during the debate. In order to set up criteria for further examination of State practice, the first issue which needs to be examined is whether a legal system requires each Minister to be answerable to the Parliament and to the public? Secondly, due to the complex nature of marine affairs, ocean governance might impinge upon the activities other public authorities. As a result, is there a need to establish a clear jurisdictional demarcation between the various ocean governance competency authorities?

Equitable and Inclusive: Good decision making needs to consider all the relevant interests, in particular, those which are in a minority in society, such as indigenous people.[132] This is illustrated by the 1992 Convention on Biological Diversity, which impose an obligation to "respect, preserve and maintain knowledge, innovations and practices of indigenous and local communities embodying traditional lifestyles …and encourage the equitable sharing of the benefits arising from the utilization of such knowledge, innovations and practices."[133]

---

[130] Andrea Ross, "The UK Approach to Delivering Sustainable Development in Government: A Case Study in Joined-up Working", Journal of Environmental Law, vol. 17 no I (2005): 31; see also Hilaire Barnett, *Constitutional and Administrative Law* 4th ed., (UK: Cavendish Publishing Limited, 2002), pp. 107; Brian Thompson, *Constitutional and Administrative Law* 3rd ed., (UK: Blackstone Press Limited, 1997), pp. 258–260.

[131] Richard A. Barnes, "Some Cautions about Integrated Oceans and coastal Management", Environmental Law Review no. 8 (2006): 251.

[132] UNDP, *Measuring Democratic Governance—A Framework for Selecting Pro-Poor and Gender Sensitive Indicators* (May 2006), pp. 7; the Report of the United Nations Committee for Development Planning, *Poverty Alleviation and Sustainable Development: Goals in Conflict?*, supra note 42, pp. 62–63; Thomas M. Franck, "Legitimacy in the International System", supra note 52: 725–735; Peter Rogers and Alan W Hall, supra note 58: 9.

[133] Article 8 (j) of the 1992 Convention on Biological Diversity.

International law emphasizes the need for equality and the fair allocation of natural resources.[134] As the Preamble to UNCLOS states, its objective is to achieve "the realization of a just and equitable international economic order..., in particular, the special interests and needs of developing countries, whether coastal or land-locked." Furthermore, the 1998 Aarhus Convention states that it is necessary to encourage "the possibility to participate in decision-making and have access to justice in environmental matters without discrimination as to citizenship, nationality or domicile and, in the case of a legal person, without discrimination as to where it has its registered seat or an effective centre of its activities."[135] This concept should also be introduced into domestic law and implemented by decision makers.

At national level, public authorities make many decisions, most of which will have an effect on individuals or interest groups. Decisions taken should be on the basis of common agreement among all those involved in the decision making process, which is also required, in order to be in sympathy with the element of 'consensus based decision making'. The courts have to be able to recognize when a decision will be subject to the requirements of fairness and when it will not.[136] In the light of this, the element of 'rule of law' should once again be combined with the element of 'equitable and inclusive'. Equality and inclusiveness in decision-making can also be facilitated by 'public participation', thus offering a 'transparent' decision-making process. Equality and inclusiveness is, therefore, an overlapping concept. For the purpose of setting standards, the foremost element which needs to be considered is ensuring that individual rights and interests are properly respected.

Responsive: The concept of 'responsiveness' is seen to relate to the manner in which the authority responds to a matter raised. The United Nations Economic and Social Commission for Asia and the Pacific stated that "Good governance requires that institutions and processes try to serve all stakeholders within a reasonable timeframe."[137] To achieve this requirement, a specific time limit within which decisions must be reached should be dictated by law. As specified by the 1998 Aarhus Convention, the public authority shall make environmental information "available as soon as possible and at the latest within one month after the request

---

[134] The case of the Gabcikovo-Nagymaros Project (Hungary v. Slovakia), Judgment of the International Court of Justice, pp. 49–53; North Sea Continental Shelf (Federal Republic of Germany/Denmark; Federal Republic of Germany/Netherlands) (1967–1969), Judgment of 20 February 1969, pp. 1–17; 83–101; see also Chapters 23, 24, 25 of Agenda 21.

[135] Article 3 (9) of the 1998 Aarhus Convention.

[136] Mungo Deans, *Scots Public Law* (Edinburgh: T & T Clark, 1995), pp. 154.

[137] The United Nations Economic and Social Commission for Asia and the Pacific, Human Settlements, "What is Good Governance?", available from: http://www.unescap.org/huset/gg/governance.htm last visited date: 07/06/2006. See also UNDP, *Governance for Sustainable Human Development*, supra note 39, pp. 1–4; The Report of the United Nations Committee for Development Planning, *Poverty Alleviation and Sustainable Development: Goals in Conflict?*, supra note 42, pp. 62–63; Peter Rogers and Alan W Hall, supra note 58: 9; Patrick Molutsi, supra note 31: 5.

has been submitted, unless the volume and the complexity of the information justify an extension of this period up to two months after the request."[138]

Another aspect of responsiveness is that the legal system needs to respond to the current needs of the environment and to current social and economic needs. The Contracting Parties to the 1992 OSPAR Convention are required to adopt the best available techniques and best environmental practice,[139] by which to ensure that their decision-making system is responsive to the needs of individuals or the environment. Another example of this can be found in the US Ocean Act 2000 which states that within 90 days of receipt of the commission report, the President is required to provide Congress with a statement of proposals to implement or respond to "the commission's recommendations for a coordinated, comprehensive, and long-range national policy for the responsible use and stewardship of ocean and coastal resources ..."[140]

As discussed earlier, the elements of 'responsive', 'rule of law' and 'transparent' are interrelating in nature. In terms of setting up criteria for further examination of State practice, the following questions have to be considered: Does legislation impose a legal duty on public authorities to make decisions within a specific time scale? Does the legal system provide a mechanism to respond to the current needs of the general public/environment?

Coherent: The element of 'coherent' relates to the processes and institutions' producing results that meet the needs of society, while making the best use of the resources at their disposal.[141] As the Preamble of UNCLOS states, it is the United Nations' intention to encourage all states to try to carry out "equitable and efficient utilization of their resources." This, therefore, requires a certain degree of co-operation and co-ordination amongst the participants. This call has been echoed by the Gabcikovo-Nagymaros Project (Hungary v. Slovakia) case, in which the International Court of Justice (ICJ), for the first time, indicated that,

"..., new norms and standards have been developed, set forth in a great number of instruments during the last two decades. Such new norms have to be taken into consideration, and such new standards given proper weight, not only when states contemplate new activities but also when continuing with activities begun in the

---

[138] Article 4 (2) of the 1998 Aarhus Convention.

[139] Article 2.2 of the 1992 OSPAR Convention.

[140] Section 4 (a) of the Oceans Act 2000.

[141] United Nations Economic and Social Commission for Asia and the Pacific, Human Settlement, "What is Good Governance?" Available from: http://www.unescap.org/huset/gg/governance.htm last visited date: 07/06/2006. See also The Report of the United Nations Committee for Development Planning, *Poverty Alleviation and Sustainable Development: Goals in Conflict?*, supra note 42, pp. 62–63; Commission of the European Communities, *White Paper on European Governance—Enhancing Democracy in the European Union* (October 2000), pp. 4; Thomas M. Franck, "Legitimacy in the International System", supra note 52: 712–759; Thomas M. Franck, "The Power of Legitimacy and the Legitimacy of Power: International Law in an Ago of Power Disequilibrium", supra note 52: 93.

past. This need to reconcile economic development with protection of the environment is aptly expressed in the concept of sustainable development."[142]

With the aim of achieving sustainable development, the ICJ states that new measures need to be taken in conjunction with scientific information, thus, those new measures or approaches taken need to be consistently applied to activities begun in the past and to those in the future. The call for the 'coherence' in good governance can, therefore, indicate a linkage between concept and reality. Furthermore, the elements of 'coherent' and 'responsive' are interrelated, as new measures or approaches taken should be able to respond to the changing needs of the general public and the environment.

As pointed out by the independent commission on the World's Ocean, the fragmentation of ocean law and institutions at both national and international levels is a major obstacle to effective ocean governance. While recognizing that there was no single, optimal model that all states should adopt, in order to encourage coherency, the commission strongly emphasized the need for states to "establish at a high governmental level an appropriate policy and coordinating mechanism, to set and review national goals for ocean affairs."[143] The aforesaid body should be able to reconcile policies and programs in order to promote an integrated approach to ocean governance. At present, efforts are being made to develop new policies and institutional arrangements for effective and efficient ocean governance and to establish integrated ocean policies, in order to overcome the problems of the sectoral based approaches to the oceans.[144]

The above mentioned approach has been implemented by a number of coastal states. In the US, the Oceans Act was introduced in 2000 to establish a new commission on ocean policy, whose remit was to make recommendations for "coordinated and comprehensive national ocean policy".[145] In Canada, the Minister of Fisheries and Oceans, "in collaboration with other ministers, boards and agencies of the Government of Canada, with provincial and territorial governments and with affected aboriginal organizations, coastal communities and other persons and bodies, including those bodies established under land claims agreements, shall lead and facilitate the development and implementation of a national strategy for the management of estuarine, coastal and marine ecosystems in waters ..."[146] This suggests that in order to achieve better coherence, any decision taken needs to be 'equitable and inclusive'.

---

[142] ICJ Report (1997), 7, at para. 140. See also US-Import Prohibition of Certain Shrimp and Shrimp Products (Shrimp-Turtle case), WTO Appellate Body (1998) WT/DS58/AB/R.

[143] Independent World Commission on the Ocean, *The Ocean: Our Future* (UK: Cambridge University Press, 1998), pp. 152.

[144] Lawrence Juda, "Changing National Approaches to Ocean Governance: The United States, Canada, and Australia", Ocean Development and International Law, no. 34 (2003): 164–165.

[145] U.S. Commission on Ocean Policy official website: http://www.oceancommission.gov/documents/oceanact.html last visited date: 5/7/2006.

[146] Section 29 of the Oceans Act 1996.

In Australia, the National Oceans Office is the main government agency responsible for the implementation of the country's oceans policy. Within the National Oceans Office, the Marine Conservation Branch has a range of responsibilities for marine environment policy development and implementation. This branch has primary responsibility for the development and implementation of integrated oceans management and for Marine Protected Areas. The branch also coordinates marine policy throughout the department and provides high level policy advice on a range of marine issues. The Marine Environment Branch undertakes activities at regional, national and international levels on a range of issues, including responsibility for protected and migratory marine species and managing Australia's obligations under UNCLOS.[147]

As discussed earlier, the element of 'coherent' functionally overlaps with other elements such as 'responsive', 'equitable and inclusive'. In order to clarify the criteria for further examination of State practice, the following issue should be considered: from an organizational aspect, is there a high level institution coordinating the various public authorities, in order to facilitate a comprehensive ocean governance policy?

1.4 Conclusion

As already discussed, the current situation as regards policy development in relation to oceans governance at international level is that there are no clear mechanisms or policy approaches in place to foster cooperation and coordination in a way that could comprehensively and effectively address the conservation of marine ecosystems. Given this situation, it is very difficult to implement good ocean governance policy at the national, regional and global levels.

While there is a great deal of literature which touches upon the issue of 'good governance', what constitutes the elements of good governance varies depending upon the views of different authors or institutions. This chapter reviews literature in relation to good governance and establishes which of those elements which contributing to good governance are most commonly accepted as being necessary to the concept.

It can be concluded that good governance is a subjective term that assumes that the goals and benchmarks for what is good have been predefined by the stakeholders. This chapter, however, suggests eight elements which are the more commonly accepted as elements which contribute to the concept of good governance, these being, the *rule of law, participatory, transparency, consensus based decision making, accountable, equitable and inclusive, responsive* and *coherent*.

---

[147] Department of Environment, Water, Heritage and the Arts, Australian Government, available from: http://www.environment.gov.au/coasts/oceans-policy/index.html last visited date: 08/07/2008.

The paper also provides examples from international treaty and State practice for each element of good governance. The aim of this exercise is to distinguish each individual element of good governance for the purposes of establishing criteria for a further examination of practice. The outcome also provides an illustration as to what the elements of good governance may look like in international treaty and State practice. It can be concluded that each element of good governance is, to some extent, supported by international treaty and State practice.

Furthermore, the elements of good governance play the role of indicators.[148] The information regarding jurisdiction should help to illustrate where the problems exists and provide the basic direction for further examination of practice. Problems may occur, however, when not all of the stakeholders have been identified or have not been involved in the process of 'setting up the rules of the game'. Problems may also exist when the perspective is limited by institutional barriers and interests or through a lack of information provided by government.[149]

Based on the aforesaid, this research suggests that good ocean governance needs national and local governmental interdisciplinary coordination and cooperation. This may be achieved by setting up a highest level organization to tackle this issue as a whole or establishing a cross-cutting policy decision-making mechanism.[150] Perpendicularly, national policies need the participation of all bureaucracy levels, as well as public and the private enterprises, whereas, horizontally, there is a need for the integrated involvement of the various government departments.

---

[148] Lynda M. Warren, "Sustainable Development and Governance", Environmental Law Review 5 (2003): 77–85; Scottish Executive, *Choosing Our Future—Fragile Handle with Care*, (Edinburgh: December 2005), para. 14.21–23, 77–78.

[149] Don Walsh, "Some Thoughts on National Ocean Policy: The Critical Issue", San Diego Law Review 13 (1975–76):609; William M. Eichbaum, "A Comprehensive Ocean Governance System in the United State—Testimony", see http://www.oceancommission.gov/meetings/sep24_25_02/eichbaum_testimony.pdf Last visited date: 07/06/2007; Sue Nichols, David Monahan and Michael Sutherland, "Good Governance of Canada's Offshore and Coastal Zone: Towards an Understanding of the Marine Boundary Issues", see http://gge.unb.ca/Research/OceanGov/documents/geomatica.pdf last visited date: 12/4/2004, website no longer exists, copy in file.

[150] As Walsh identifies "Two possible approaches are as follow: first, to provide continuous, evaluated policy information and recommendations to the highest levels of the executive and legislative branches of Government. The second is to unify and harmonize the various mission-related activities of the many local organizations which have marine science and ocean affairs programs". For detail please see Don Walsh, supra note 149: 619. The same view please see Andrea Ross-Robertson, "Is the Environment Getting Squeezed out of Sustainable Development?", Public Law (Summer, 2003): 249–259. For different view please see Fidel V. Ramos (2001), supra note 48: 16.

Chapter 2
Good Ocean Governance and International Law

Abstract At the present, the concept of good ocean governance is articulated by literature only. This chapter adopted eight elements of good governance as an analytical framework, namely, the rule of law, participatory, transparency, consensus based decision making, accountability, equity and inclusiveness, responsiveness and coherence. The chapter also provides evidence from international treaty practice to support each element of good ocean governance. In summary, the elements of good ocean governance are partially supported by international treaty practice but are not yet receiving universal acceptance.

Keywords Good ocean governance · UNCLOS · 1998 Aarhus convention · 1972 London convention · MARPOL 73/78 · 1992 OSPAR convention

2.1 Introduction

Although there has been a world-wide effort to develop an effective ocean governance mechanism,[1] to date, there are no specific mechanisms or policy approaches in place, with which to encourage cooperation and coordinated action.

---

[1] General Assembly, United Nations, *Oceans and the Law of the Sea—Report of the Secretary-General*, A/61/63, 9 March 2006, pp. 76–78.

'Good governance', is considered to be a positive and constructive element of sustainable development.[2] It is an open decision-making process which should involve public participation, the release of environmental information and access to environmental justice.[3]

Eight elements were acknowledged as being the elements of good governance, namely, the rule of law, participatory, transparency, consensus based decision making, accountability, equity and inclusiveness, responsiveness and coherence.[4] It is proposed that good ocean governance may be achieved by means of major international law of the sea treaties, through the operation of global and regional

---

[2] Konrad Ginther and Paul J. I. M. de Waart, "Sustainable Development as a Matter of Good Governance: An Introductory View", in Konrad Ginther, Erik Denters and Paul J. I. M. de Waart (ed), *Sustainable Development and Good Governance*, (London: Martinus Nijhoff Publishers, 1995), pp. 9.

[3] See Preamble of the Convention on Access to Information, Public Participation in Decision-Making and Access to Justice in Environmental Matters (Aarhus) 25 June 1998, came into force on 30th October 2001, 38 ILM 517 (1999).

[4] Yen-Chiang Chang, "Good Ocean Governance", in *Ocean Yearbook*, vol. 23, 2009, pp. 89–118; Francis N. Botchway, "Good Governance: The Old, The New, The Principle, and The Elements", Florida Journal of International Law, no.13, 2001, pp. 180–183; The European Commission, *Communication on Governance and Development*, October 2003, COM (03) 615; The World Bank, *Governance and* Development, Washington, 1992, p. 1; UNDP, *Governance Indicators: A User's Guide*, p. 3, available from: http://www.undp.org/oslocentre/docs04/UserGuide.pdf last visited date: 21/12/2007; Asian Development Bank, *Governance: Sound Development Management*, August 1999, p. 3; Tony Bovaird and Elke Löffler, "Evaluating the Quality of Public Governance: Indicators, Models and Methodologies", International Review of Administrative Sciences, 69 no. 3, 2003, p. 316; Robert I. Rotberg, "Strengthening Governance: Ranking Countries Would Help", Washington Quarterly, 28:1, Winter 2004–5, p. 71; Peter Rogers and Alan W Hall, *Effective Water Governance*, Global Water Partnership Technical Committee, The Background Papers, No. 7, February 2003, p. 4; Andrew Allan and Patricia Wouters, 'What Role for Water Law in the Emerging "Good Governance" Debate?', available from www.dundee.ac.uk/water/Documents/Conferences/2004/woutersGovernancearticlefor AWRA.pdf. last visited date: 18/03/2008; Patrick Molutsi, "Tracking Progress in Democracy and Governance Around the World: Lessons and Methods", available from: http://unpan1.un.org/intradoc/groups/public/documents/un/unpan005784.pdf last visited date: 22/12/2007; Carlos Santiso, "Good Governance and Aid Effectiveness: The World Bank and Conditionality", The Georgetown Public Policy Review, vol. 7 no. 1, Fall 2001, p. 2; Daniel Kaufmann, Aart Kraay, Pablo Zoido-Lobatón, *Aggregating Governance Indicators*, The World Bank Policy Research Working Paper 2195, October 1999, pp. 1–2, 5–7; Commission of the European Communities, *White Paper on European Governance—Enhancing Democracy in the European Union*, Brussels, 11th October 2002, SEC (2000) 1547/7 final, pp. 4; see also European Union, *Governance in the European Union: A White Paper—Enhancing Democracy in the European Union*, COM (2001) 428 final; European Commission, *Report from the Commission on European Governance (European Commission, 2003)*; see also European Commission, *Toward a Reinforced Culture of Consultation and Dialogue—General Principles and Minimum Standards for Consultation of Interested Parties*, COM (2002) 704 final.

organisations and through national ocean governance efforts.[5] Based on the above assumptions, it is therefore necessary to discuss how good ocean governance has been incorporated into international treaty practice.

This chapter aims to provide examples from the existing international law of the sea, to illustrate how good governance has been incorporated into international treaty practice. In order to achieve this objective, the author adopts a two-stage approach. Firstly, this chapter selects a number of international treaties that are relevant to ocean governance and indicates the reasons for the selection. Then, the chapter will focus on the elements of good governance and their relationship with international treaties. The purpose of this exercise is to find examples from international treaty practice of each element of good ocean governance. The overall outcome will provide legal evidence or support to the elements of good ocean governance.

In brief, the outcome of this analysis is that although the concept of good ocean governance is not, as yet, expressly or universally accepted by international law, each element of good ocean governance has, at least to some extent, been addressed by international law treaty practice. A State that is complying with its treaty obligations will, in practice, be applying the elements of good governance, even although this may not be made explicit at international level. It is, therefore, not unreasonable to suggest that there is a legal requirement to introduce the concept of good ocean governance at the domestic level.

2.2 The Legal Governance Framework for the Oceans

The 1982 United Nations Convention on the Law of the Sea (UNCLOS)[6] aims to establish a legal order for the seas and oceans, which will facilitate international communication and promote the peaceful use of the seas and oceans, as well as providing for the equitable and efficient utilisation of their resources and the conservation of their living resources.[7] Part XII of UNCLOS specifically sets out obligations for the protection and preservation of the marine environment, the said obligations also providing an implementation framework for ocean governance and therefore, examples in relation to the elements of good governance from this part of treaty will be provided.

There are four types of marine pollution detailed by UNCLOS, namely, pollution from dumping, pollution from land-based sources, including through the atmosphere, pollution from vessels and pollution from seabed activities.[8] This

---

[5] Robert L. Friedeim, "Ocean Governance at the Millennium: Where We Have Been—Where We Should Go", Ocean & Coastal Management, vol. 42, 1999, p. 747.

[6] UNCLOS, 10th December 1982, came into force 16th November 1994, 21 ILM 1261. At the time of writing, there are 155 parties to UNCLOS.

[7] Preamble of UNCLOS.

[8] UNCLOS Article 194 (3) (a)–(d).

study will, based on the foregoing, examine the 1972 Convention on the Preservation of Marine Pollution by Dumping of Waste and Other Matter[9] (1972 London Convention)[10] and its 1996 Protocol,[11] the 1992 Convention for the Protection of Marine Environment of the North-East Atlantic[12] (1992 OSPAR Convention),[13] the International Convention for the Prevention of Pollution from Ships, as modified by the Protocol of 1978,[14] usually referred to as the MARPOL 73/78.[15]

The 1992 Convention on Biological Diversity,[16] together with the 1995 United Nations Fish Stocks Agreement,[17] are of central importance as regards the governance of marine living resources. The 1992 Convention on Biological Diversity was the first treaty to provide a legal framework for biodiversity conservation and it established three main goals: the conservation of biological diversity, the sustainable use of its components and the fair and equitable sharing of the benefits arising from the use of genetic resources.[18] The 1995 United Nations Fish Stocks Agreement aims to ensure the long-term conservation and sustainable use of straddling fish stocks and highly migratory fish stocks.[19] In particular, it requires States to co-operate so that there is compatibility between national and high seas

---

[9] Convention on the Preservation of Marine Pollution by Dumping of Waste and Other Matter, (London, Mexico City, Moscow, Washington DC) 29th December 1972, came into force on 30th August 1975; 1046 UNTS 120.

[10] At the time of writing, there are 89 parties to the 1972 London Convention.

[11] The 1996 Protocol to the Convention on the Prevention of Marine Pollution by Dumping of Wastes and Other Matter (London), 7th November 1996, came into force on 24th March 2006, 36 ILM 1.

[12] The 1992 Convention for the Protection of Marine Environment of the North-East Atlantic (Paris), 22nd September 1992, came into force on 25th March 1998, 23 LOSB 32.

[13] At the time of writing, there are 16 parties to the 1992 OSPAR Convention.

[14] Protocol Relating to the Convention for the Prevention of Pollution from Ships (London), 17th February 1978, came into force on 2nd October 1983, 17 ILM 546.

[15] Apart from the MARPOL 73/78 there is another convention relating to the pollution from vessels, which is the International Convention on Civil Liability for Oil Pollution Damage 1992 but this convention was adopted to ensure that adequate compensation is available to persons who suffer oil pollution damage resulting from maritime casualties involving oil-carrying ships and places the liability for such damage on the owner of the ship from which the polluting oil escaped or was discharged. Since the notion of good governance is more focused on the decision-making process of public authorities, civil liability is therefore beyond the scope of this study.

[16] Convention on Biological Diversity (Rio de Janeiro), 5th June 1992, came into force on 29th December 1993, 31 ILM 822 (1992). At time of writing, there are 190 parties to the 1992 Convention on Biological Diversity.

[17] Agreement for the Implementation of the Provisions of the United Nations Convention on the Law of the Sea of 10 December 1982 Relating to the Conservation and management of Straddling Fish Stocks and Highly Migratory Fish Stocks (New York), 4th December 1995, came into force on 11th December 2001, 34 ILM 1542 (1995). At time of writing, there are 66 parties to the 1995 United Nations Fish Stocks Agreement.

[18] Preamble of the 1992 Convention on Biological Diversity.

[19] Article 2 of the 1995 United Nations Fish Stocks Agreement.

measures.[20] This chapter will, therefore, examine their views on good governance and where appropriate, examples from other international environmental treaties will be provided.

Academics and others, increasingly aware of the need for a broader form of ocean governance, began calling for more comprehensive, better integrated approaches.[21] This need began to be echoed at international conferences and in declarations such as the Rio Declaration.[22] The author also focuses on the way in which the concept of good governance and its earlier manifestations have been reflected in international agreements, with special attention being devoted to the pronouncements related to the United Nations Conference on the Human Environment held in Stockholm, in June 1972,[23] the United Nations Conference on Environment and Development in Rio de Janeiro, Brazil, in June 1992[24] and the United Nations World Summit on Sustainable Development in Johannesburg, South Africa, in September 2002.[25] In the thirty-year time span represented by these conferences, there can be seen the gradual maturation of the concept of good governance as a generally accepted concept, which is articulated progressively in the national decision making processes.

As was agreed at the Rio Conference, the protection of the environment and social and economic development are fundamental to sustainable development. To achieve this development, the international community adopted the global programme entitled Agenda 21,[26] which deals with marine issues. In particular, in Chapter 17, it states that its objective is: *Protection of the oceans, all kinds of seas, including enclosed and semi-enclosed seas, and coastal areas and the protection,*

---

[20] UN Oceans and Law of the Sea website: http://www.un.org/Depts/los/convention_agreements/convention_overview_fish_stocks.htm last visited date: 27/02/2007.

[21] Mariam Sara Repetto, "Towards and Ocean Governance Framework and National Ocean Policy for Peru", The Nippon Foundation of Japan Fellow, 2005, available from: http://www.un.org/depts/los/nippon/unnff_programme_home/fellows_pages/fellows_papers/repetto_0506_peru.pdf last visited date: 21/07/2008, p. 8–9; Seoung-Yong Hong and Young-Tae Chang, "Integrated Coastal Management and the Advent of New Ocean Governance in Korea: Strategies for Increasing the Probability of Success", The International Journal of Marine and Coastal Law, no. 2, 1997, p. 142; General Assembly of the United Nations, *Oceans and Law of the Sea—Report of the Secretary-General*, A/60/63, 4 March 2005, p. 80.

[22] United Nations, *Report of the United Nations Conference on Environment and Development*, Annex I, Rio Declaration on Environment and Development, UN Doc A/CONF.15/26 (Vol. I) (1992) (Rio Declaration); Gillian D. Triggs, *International Law: Contemporary Principles and Practices* (Australia:LexisNexis Butterworths, 2006), pp. 803–804; Philippe Sands, *Principles of International Environmental Law* 2nd ed. (UK: Cambridge University Press, 2003), pp. 4–5.

[23] Stockholm Declaration of the United Nations Conference on the Human Environment, UN Doc A/CONF.48/14 (1972) (XXVII), UNGAOR, 27th sess, 2112th plen mtg, UN Doc A/RES/2994 (XXVII).

[24] United Nations publication, Sales No. E.73.II.A.14 and corrigendum, Chap. I.

[25] Adopted at the 17th plenary meeting of the World Summit on Sustainable Development, on 4th September 2002, see Chap. VIII of the Summit Report.

[26] United Nations, *Report of World Summit on Sustainable Development*, Johannesburg, South Africa, 26th August–4th September 2002, ISBN 92-1-104521-5, 8.

rational use and development of their living resources. The aforementioned 'soft law' consists of non-treaty obligations, however, it incorporates the fundamental components of the international legal order. Following on from the aforementioned, the subsequent section will discuss the extent to which the elements of good governance are being incorporated into international treaty practice.

2.3 International Law and Good Ocean Governance

International law and institutions serve as the main framework for international cooperation and collaboration between members of the international community in their efforts to protect the local, regional and global marine environment. The subsequent section will provide examples of what elements of good ocean governance may look like in practice. The aforesaid elements of good governance being: the rule of law, participatory, transparency, consensus based decision making, accountability, equity and inclusiveness, responsiveness and coherence. The aim is to provide an insight into the sorts of provisions one would anticipate States may adopt or respond to, in ensuring that they implement good ocean governance.

2.3.1 The Rule of Law

The rule of law emphasises that all laws have to be published via appropriate media, equally and fairly administered and effectively enforced. It is also important to ensure that decision makers and administrators are bound to follow the rule of law when making decisions. The subsequent part will provide examples of the rule of law from international treaty practice.

Under UNCLOS, States are asked to adopt laws, regulations, measures, rules, standards, recommended practices and procedures to prevent, reduce and control pollution of the marine environment from varied ocean use activities.[27] The content of these national instruments shall be "no less effective in preventing, reducing and controlling such pollution than the global rules and standards."[28] The evaluation of enforcement with respect to these national instruments to marine pollution control activities is through competent international organisations or diplomatic conference.[29] Nevertheless, UNCLOS leaves room for States to decide "the best practicable means at their disposal and in accordance with their

---

[27] Article 207.1 and 207.5; 208.1 and 208.2; 209.2; 210.1, 2 and 3; 211.2; 212.1 and 212.2 of UNCLOS.

[28] Article 208.3; 209.2; 210.6 and 211.2 of UNCLOS.

[29] Article 213, 214, 216, 217 of UNCLOS.

capabilities"[30] to prevent and control marine pollution. In brief, UNCLOS establishes State responsibility not to cause damage to the marine environment and this has commonly been accepted as customary international law.[31] States are, therefore, required to follow the rule established by UNCLOS and transpose this into national law or laws. The content of the said national law or laws must be as clear as possible and cover all aspects of concern regarding ocean governance.[32]

The ethos of following the rule of law is evidenced in the MARPOL 73/78, as the Parties to the MARPOL 73/78 shall "undertake to give effect to the provisions of the present Convention and those Annexes thereto by which they are bound, in order to prevent the pollution of the marine environment by the discharge of harmful substances or effluents containing such substances in contravention of the Convention."[33] To achieve these objectives, the Parties may promulgate domestic law to provide certificates and special rules on the inspection of ships.[34] Any Parties may deny "a foreign ship entry to the ports or off-shore terminals under its jurisdiction or take any action against such a ship for the reason that the ship does not comply with the provisions of the present Convention, the Parties shall immediately inform the consul or diplomatic representative of the Party whose flag the ship is entitled to fly...."[35] It is necessary for Parties to ensure that no more favourable treatment is given to the ships of non-Parties to the Convention.[36] This reading indicates that the Contracting Parties are obliged to publish law in an appropriate way under the MARPOL 73/78. In addition, these laws should be equally and fairly administered, regardless of whether dealing with Parties or non-Parties to the Convention. The text of laws, orders, decrees and regulations and other instruments which have been promulgated on the various matters within the scope of the MARPOL 73/78 are subject to submission to the Inter-Governmental Maritime Consultative Organisation.[37](IMCO) The aforesaid Organisation is entitled to monitor the implementation of the MARPOL 73/78. The competent authorities of the Parties are, therefore, bound to follow the rule of law when making decisions.

Under the 1992 Convention on Biological Diversity, each Contracting Party shall "Regulate or manage biological resources important for the conservation of biological diversity whether within or outside protected areas, with a view to ensuring their conservation and sustainable use."[38] The Contracting Parties shall

---

[30] Article 194.1 of UNCLOS.

[31] *Nuclear Tests* case (*Australia* v. *France*) (1974) ICJ Reports 253 at 389; *United States* v. *Canada*, 3 RIAA 1907 (1941).

[32] Articles 194.3 and 207.5 of UNCLOS.

[33] Article 1.1 of the MARPOL 73/78.

[34] Article 5 of the MARPOL 73/78.

[35] Article 5.3 of the MARPOL 73/78.

[36] Articles 5.3 of the MARPOL 73/78.

[37] Article 11.1 (a) of the MARPOL 73/78.

[38] Article 8 (c) of the 1992 Convention on Biological Diversity.

also "develop or maintain necessary legislation and/or other regulatory provisions for the protection of threatened species and populations."[39] Legislative, administrative or policy measures shall be taken to ensure that international law is followed[40] in "a fair and equitable way."[41] The Contracting Party is, therefore, obliged to publish law consistent with the 1992 Convention on Biological Diversity and the said law should be implemented equitably and fairly.

Finally, the clear and equitable rule of law at national level is essential. As stated by the 1992 Rio Declaration, "States shall enact effective environmental legislation. Environmental standards, management objectives and priorities should reflect the environmental and development context to which they apply."[42] They also need to "develop national law regarding liability and compensation for the victims of pollution and other environmental damage."[43] These measures should not "constitute a means of arbitrary or unjustifiable discrimination or a disguised restriction."[44] The 1992 Rio Declaration establishes a need to enact environmental law in order to protect the environment and the said environmental law should be implemented in a fair and equitable matter.

From the above, international treaty practice has set out examples to publish laws in an appropriate way. International legal instruments also emphasise the need that the law should be equally and fairly administered and effectively enforced. At the domestic level, decision makers and administrators are bound to follow the rule of law, when making decisions.

2.3.2 Participatory

The following part will provide examples from international treaty practice as illustrations of public participation. This research will also consider in what form the participatory activity is organised.

The 1998 Convention on Access to Information, Public Participation in Decision-Making and Access to Justice in Environmental Matters[45] (1998 Aarhus Convention), describes public participation as: designed to give members of the public "the opportunity to express their concerns and enable[ing] public

---

[39] Article 8 (k) of the 1992 Convention on Biological Diversity.

[40] Article 16 (3) of the 1992 Convention on Biological Diversity.

[41] Article 15 (7) of the 1992 Convention on Biological Diversity.

[42] Principle 11 of the 1992 Rio Declaration.

[43] Principle 13 of the 1992 Rio Declaration.

[44] Principle 12 of the 1992 Rio Declaration.

[45] Convention on Access to Information, Public Participation in Decision-Making and Access to Justice in Environmental Matters (Aarhus) 25th June 1998, came into force on 30th October 2001, 38 ILM 517 (1999).

authorities to take due account of such concerns."[46] The 1998 Aarhus Convention also provides detailed and clear obligations on States to include in their decision-making processes the concept of public participation. These obligations include: public participation in decisions on specific activities[47]; public participation concerning plans, programmes and policies relating to the environment[48] and public participation during the preparation of executive regulation and/or generally applicable legally binding normative instruments.[49] The contracting parties should, therefore, fix the time-frames appropriately to "promote effective public participation at an appropriate stage, and while options are still open, during the preparation by public authorities of executive regulations and other generally applicable legally binding rules that may have a significant effect on the environment."[50] It is also the contracting parties' obligation to "provide for early public participation, when all options are open and effective public participation can take place."[51]

In addition, the 1992 Convention on Biological Diversity imposes obligations to allow public participation when introducing "appropriate procedures requiring environmental impact assessment of its proposed projects that are likely to have significant adverse effects on biological diversity with a view to avoiding or minimizing such effects."[52] Public participation in the authorities' decision-making procedure is, therefore, no longer merely a political theme, it is a legally binding obligation.

Apart from 'hard law' sources, the 1992 Rio Declaration also indicates that the participation of the public is necessary. As the 1992 Rio Declaration states, "Women have a vital role in environmental management and development. Their full participation is therefore essential to achieve sustainable development"[53]; "the creativity, ideals and courage of the youth of the world should be mobilized to forge a global partnership"[54]; "Indigenous people and their communities and other local communities have a vital role in environmental management and development because of their knowledge and traditional practices. States should recognize and duly support their identity, culture and interests and enable their effective participation"[55] and finally, "States and people shall cooperate in good faith and in a spirit of partnership in the fulfilment of the principles embodied in this

---

[46] Preamble of the 1998 Aarhus Convention.

[47] Article 6 of the 1998 Aarhus Convention.

[48] Article 7 of the 1998 Aarhus Convention.

[49] Article 8 of the 1998 Aarhus Convention.

[50] Article 8 (a) of the 1998 Aarhus Convention.

[51] Article 6 (4) of the 1998 Aarhus Convention.

[52] Article 14 (a) of the 1992 Convention on Biological Diversity.

[53] Principle 20 of the 1992 Rio Declaration.

[54] Principle 21 of the 1992 Rio Declaration.

[55] Principle 22 of the 1992 Rio Declaration.

Declaration and in the further development of international law in the field of sustainable development."[56]

Chapter 17 of Agenda 21 provides a joint approach to access, where ever possible, "for concerned individuals, groups and organisations to relevant information and opportunities for consultation and participation in planning and decision-making at appropriate levels is necessary."[57] It is also important to consult "on coastal and marine issues with local administrations, the business community, the academic sector, resource user groups and the general public."[58] Without an appropriate assessment mechanism, however, the decision-making processes may not be deemed to be complete. As a result, it is imperative "to review the existing institutional arrangements and identify and undertake appropriate institutional reforms essential to the effective implementation of sustainable development plans, including inter-sectoral coordination and community participation in the planning process."[59]

The 2002 Johannesburg Declaration emphasises the link to equitable public participation within the decision-making processes. It requires a "long-term perspective and broad-based participation in policy formulation, decision-making and implementation at all levels."[60] The continuous striving for stable partnerships involving all major groups is an important part of public participation. "The promotion of dialogue and cooperation among the world's civilizations and peoples, irrespective of race, disabilities, religion, language, culture or tradition"[61] are, therefore, relevant and necessary requirements.

It can be seen that these international legal instruments provide an obligation to ensure that there is public participation, which should lead to open decision-making processes. Although the precise form of public participation is not yet clear from the reading of treaties, this could be achieved through consultation.[62] Following the above discussion, the public should be given the opportunity to be involved in decisions, concerns, plans, programmes and policy relating to the environment, as well as in the preparation of executive regulations and other legal instruments.

---

[56] Principle 27 of the 1992 Rio Declaration.

[57] Paragraph 17.5 (f) of the Agenda 21.

[58] Paragraph 17.17 (b) of the Agenda 21.

[59] Paragraph 17.128 of the Agenda 21.

[60] Paragraph 26 of the 2002 Johannesburg Declaration.

[61] Paragraph 17 of the 2002 Johannesburg Declaration.

[62] Paragraph 17.17 (b) of the Agenda 21.

2.3.3 Transparency

The element of transparency emphasises the need for the public authorities to release environmental information, including information on decision making. It is also important to consider in what form is the environmental information should be presented. The subsequent part will provide examples to illustrate what the element of transparency may look like, from the viewpoint of international treaty practice.

UNCLOS sets out substantive rules and standards to facilitate transparent decision making including: notifying of imminent or actual damage[63]; developing contingency plans against pollution[64]; promoting the studies, research programmes and exchange of information and data, as well as providing the scientific criteria for regulations and[65] monitoring of the risks or effects of pollution, all of which should be included in an environmental assessment report.[66] This would suggest that UNCLOS imposes obligations on States to pro-actively release or exchange environmental information, which should be presented in the form of an environmental assessment report. States should provide the aforesaid reports to the competent international organisations, which, in turn, should make the reports available to all States.[67] In this instance, transparency is to other States but not to the public, as there is no reference to releasing the minutes of meetings to the public. It is also a limited form of transparency in relation to States, as only specific information is released but not information on decision making. Nonetheless, it provides an example of what transparency means in practice.

Article 12 of the 1995 United Nations Fish Stocks Agreement emphasises the need to "provide for transparency in the decision-making process and other activities of sub-regional and regional fisheries management organizations and arrangements."[68] To facilitate the aforesaid objective, States shall establish, "a national record of fishing vessels authorized to fish on the high seas and provision of access to the information contained in that record on request by directly interested States, taking into account any national laws of the flag State regarding release of such information."[69] In this case, transparency is only to other States and there is no clear reference to releasing the minutes of meetings to the public. It also relates to transparency in international decision making, not transparency at the national level. Nonetheless, it still provides an example of what transparency means in practice.

---

[63] Article 198 of UNCLOS.
[64] Article 199 of UNCLOS.
[65] Article 200 and 201 of UNCLOS.
[66] Article 204, 205 and 206 of UNCLOS.
[67] Article 205 of UNCLOS.
[68] Article 24.2 (b) of the 1995 United Nations Fish Stocks Agreement.
[69] Article 18.3 (c) of the 1995 United Nations Fish Stocks Agreement.

Further provisional duties regarding the transparency approach in good governance were illustrated by the 1998 Aarhus Convention. In accordance with Article 3 (1) of the 1998 Aarhus Convention, the contracting parties, "shall take the necessary legislative, regulatory and other measures, including measures to achieve compatibility between the provisions implementing the information, public participation and access-to-justice provisions in this Convention, as well as proper enforcement measures, to establish and maintain a clear, transparent and consistent framework to implement the provisions of this Convention." The contracting parties are also responsible, "to ensure that officials and authorities assist and provide guidance to the public in seeking access to information, in facilitating participation in decision-making and in seeking access to justice in environmental matters."[70] The authorities must also make requested information available and that in the case of refusing a public application for environmental information, the contracting parties shall take "into account the public interest served by disclosure and taking into account whether the information requested relates to emissions into the environment."[71]

It is also true that, within the framework of national legislation, it is the public authorities' duty to possess and update environmental information and establish a mandatory system to provide all information without delay to members of the public who may be affected.[72] For the purposes of disseminating environmental information, such information is progressively becoming available from electronic databases which are easily accessible to the public through public telecommunications networks. This information should include reports on the state of the environment, texts of legislation, policies and plans and programmes on or relating to, the environment.[73] Article 6 (2) of the 1998 Aarhus Convention states that, "the public concerned shall be informed, either by public notice or individually as appropriate, early in an environmental decision-making procedure, and in an adequate, timely and effective manner." Finally, the contracting parties shall set up mechanisms to examine their legislation and policy documents to also ensure that the domestic law matches international treaties, conventions and agreements on environmental issues. An assessment body in respect of this field is therefore necessary.[74] The most important factor is that transparency is directed to the public, which is different from what is included in previous treaties. It is also important to note that the public authorities should release the minutes of meetings.[75]

The 1992 OSPAR Convention ensures that the Contracting Parties' competent authorities are required to, "make available the information in written, visual,

---

[70] Article 3 (2) of the 1998 Aarhus Convention.

[71] Article 4 (4) of the 1998 Aarhus Convention.

[72] Article 5 (1) (2) of the 1998 Aarhus Convention.

[73] Article 5 (3) (4) (7) of the 1998 Aarhus Convention.

[74] Article 5 (5) (6) (8)–(10) of the 1998 Aarhus Convention.

[75] Article 4, 5 of the 1998 Aarhus Convention.

aural or data-base form on the state of the maritime area, on activities or measures adversely affecting or likely to affect it and on activities or measures introduced in accordance with the Convention."[76] The aforesaid information should be available, "in response to any reasonable request, without that person's having to prove an interest, without unreasonable charges."[77] In addition, the Contracting Parties shall publish, at regular intervals, joint assessments of the quality status of the marine environment and of its development. The aforementioned assessments should include, "both an evaluation of the effectiveness of the measures taken and planned for the protection of the marine environment and the identification of priorities for action."[78] Once again, transparency is to other States but not to the public and there is no clear reference referring to the release of the minutes of meetings to the public. More specifically, information released is focused on specific types but is not particularly transparent in relation to the decision making. Once again, however, the international provisions provide an example of the types of activity we might anticipate seeing at the national level.

At national level, as stated by the 1992 Rio Declaration, "each individual shall have appropriate access to information concerning the environment that is held by public authorities, including information on hazardous materials and activities in their communities, and the opportunity to participate in decision-making processes. States shall facilitate and encourage public awareness and participation by making information widely available. Effective access to judicial and administrative proceedings, including redress and remedy, shall be provided."[79]

In support of the above, Chapter 17 of Agenda 21 pronounces that States should improve their capacity to collect, analyse, assess and use information for the sustainable use of resources. This environmental information should include; "environmental impacts of activities affecting the coastal and marine areas"[80]; "disposal and with due regard for their technical and scientific capacity and resources, make systematic observations on the state of the marine environment"[81]; "with the support of international organizations, whether sub-regional, regional or global, as appropriate"[82]; "individually or through bilateral and multilateral cooperation and with the support, as appropriate, of international organizations, whether sub-regional, regional or global"[83]; "use existing sub-regional and regional mechanisms, where applicable, to develop knowledge of the marine environment, exchange information, organize systematic observations and assessments, and make the most effective use of scientists, facilities and

---

[76] Article 9.2 of the 1992 OSPAR Convention.
[77] Article 9.1 of the 1992 OSPAR Convention.
[78] Article 6 of the 1992 OSPAR Convention.
[79] Principle 10 of the 1992 Rio Declaration.
[80] Paragraph 17.8 of the Agenda 21.
[81] Paragraph 17.35 of the Agenda 21.
[82] Paragraph 17.56 of the Agenda 21.
[83] Paragraph 17.86 of the Agenda 21.

equipment"[84] and there is "an obligation to ensure transparency in the use of trade measures related to the environment and to provide adequate notification of national regulations."[85] As before, while this principle refers to the exchange of information between States, it hints at the type of information that ought to be made available to the public at national level. When combined with obligations under the 1998 Aarhus Convention and elsewhere as regards making decision making transparent, the implication is that the public must be able to see what scientific understanding underpins decision making.

From the above it is clear that international treaty practice emphasises the need to release environmental information. It is recommended that environmental information should be made available from electronic databases, which are easily accessible to the public through public telecommunications networks. The aforesaid environmental information should include reports on the state of the environment, texts of legislation, policies and plans and programmes on or relating to, the environment. The 1998 Aarhus Convention does ask States to release information on decision making, whereas, other treaties point only to releasing specific types of information. Apart from the 1998 Aarhus Convention, most of international treaty practice does not require the release of minutes of meetings to the public, which may in turn weaken the performance of transparency. The current climate is moving in the direction of the adoption and implementation of the 1998 Aarhus Convention. Transparency would be further strengthened if States were obliged to release 'ALL' the minutes of meetings to the public but one may anticipate that actual practice will sit somewhere between the ideal and the position illustrated by the majority of provisions discussed above.

2.3.4 Consensus Based Decision Making

The core concern of consensus based decision-making is that no individual, official or group should be able to force their individual decisions or views on others, whether through majority voting or otherwise.[86] What does international treaty practice say about consensus based decision- making? In searching for answers to the aforesaid question, the subsequent section will provide examples to illustrate consensus based decision making from international treaty practice.

---

[84] Paragraph 17.114 of the Agenda 21.

[85] Paragraph 17.118 of the Agenda 21.

[86] National Marine Sanctuaries, Monterey Bay, 'Joint Management Plan Review Working Group—Consensus Based Decision Making', available from: www.sanctuaries.noaa.gov/jointplan/mb_docs/mb_consensus.pdf last visited date: 17/06/08; Dr. John Robert Dew, 'Consensus Based Decision Making', available from: http://bama.ua.edu/~st497/ppt/consensusbaseddecision.ppt last visited date: 17/06/2008.

The best international example of consensus based decision making is, of course, provided by the international law of trade.[87] As Article IX of the 1994 Agreement Establishing the World Trade Organisation[88] explicitly states, "The WTO shall continue the practice of decision making by consensus followed under GATT 1947." The Agreement further explains 'consensus' as having been achieved, "if no Member, present at the meeting when the decision is taken, formally objects to the proposed decision."[89] Where a decision cannot be achieved by consensus, the relevant agreements provide for decisions to be made by majority voting, often with the requirement for a qualified majority to be used.[90]

Evidence also can be found from a WTO's subsidiary organ, such as the Dispute Settlement Body (DSB). For example, Article 2.4 of the 1994 Understanding on Rules and Procedures Governing the Settlement of Disputes[91] states, "Where the rules and procedures of this Understanding provide for the DSB to take a decision, it shall do so by consensus." The aim of the DSB is to secure a positive solution to a dispute. The wish is to ensure a solution mutually acceptable to all parties to a dispute.[92] Since consensus requires the approval of the conflicting parties, these parties, not the DSB, will normally have the effective final word on the dispute topics.[93] In addition, it is important to note that for some key decisions, such as the decision on the establishment of panels,[94] the adoption of panel[95] and Appellate Body reports[96] and the authorisation of suspension of concession and

---

[87] See also Miquel I Mora, "A GATT With Teeth: Law Wins Over Politics in the Resolution of International Trade Disputes", Columbia Journal of International Law, Vol. 31 (1993–1994), pp. 142–143; Raymond Vernon, "The World Trade Organization: A new Stage in International Trade and Development", Harvard International Law Journal, vol. 36, No. 2, (1995), pp. 336–337; Steven P. Croley and John H. Jackson, "WTO Dispute Procedures, Standard of Review, and Deference to National Government", The American Journal of International Law, Vol. 90 (1996), pp. 193–213; Robert E. Hudec, Enforcing International Trade Law—The Evolution of the Modern GATT Legal System (US: Butterworth Legal Publishers, 1993), pp. 357–366.

[88] The 1994 Agreement Establishing the World Trade Organisation, (1994) 33 I.L.M. 1144 and available at http://www.wto.org.

[89] See footnote of Article IX of the Agreement Establishing the World Trade Organisation.

[90] Articles IX (1), IX (3) (a) and X of the Agreement Establishing the World Trade Organisation.

[91] The 1994 Understanding on Rules and Procedures Governing the Settlement of Disputes, (1994) 33 I.L.M. 1226 and http://www.wto.org.

[92] Article 3.7 of the 1994 Understanding on Rules and Procedures Governing the Settlement of Disputes.

[93] N. David Palmeter, Peter C. Mavroidis, Dispute Settlement in the World Trade Organization—Practice and Procedure (UK: Cambridge University Press, 2004), p. 15.

[94] Article 6.1 of the 1994 Understanding on Rules and Procedures Governing the Settlement of Disputes.

[95] Article 16.4 of the 1994 Understanding on Rules and Procedures Governing the Settlement of Disputes.

[96] Article 17.14 of the 1994 Understanding on Rules and Procedures Governing the Settlement of Disputes.

other obligations,[97] the consensus requirement is in fact a 'negative' consensus requirement.[98] The 'negative' consensus requirement means that it is deemed that the DSB will make a decision, unless there is a consensus among WTO Members not to take that decision.[99] Since there will usually be at least one Member with a controversial objective included in the panel, the adoption of the panel and/or Appellate Body reports or the authorisation on suspend concessions, it is highly unlikely that there will be a consensus not to adopt these decisions. To this end, the process prescribed for the handing of a complaint does not require a vote of the contracting parties.

The 'unanimous vote' approach is used within the Commission to the 1992 OSPAR Convention.[100] Although this is not what one generally envisages as consensus based decision making, when combined with the obligation on the Contracting Parties to harmonise their policies and strategies, it also demonstrates a good example of consensus based decision making.

Although consensus based decision making is said to be an important element of good ocean governance, there is no clear direction from international law pertaining to the marine environment as to how this can be achieved. An illustration can, however, be found in international trade law, which emphasises the importance of attempting to achieve consensus before resorting to majority voting. One might anticipate that it would be very difficult to find evidence to support the consensus based decision making at the national level. What is important, however, is that all aspects of relevant opinion should be properly respected.

2.3.5 Accountability

The subsequent part will provide examples to illustrate what international treaty practice say about accountability. Within the rules of the 1972 London Convention, the national authority is authorised to control dumping behaviour in relation to the possible harm to their citizens and marine living resources within their territory.[101] This leads to an obligation to ensure marine dumping behaviour will not cause any possible harm and if it does, it is also the States' responsibility to tackle or control the pollution before it causes further damage.[102] Furthermore,

---

[97] Article 22.6 of the 1994 Understanding on Rules and Procedures Governing the Settlement of Disputes.

[98] Other decisions of the DSB such as the appointment of the Members of the Appellate Body are taken by 'normal' consensus.

[99] Peter van den Bossche, *The Law and Policy of the World Trade Organization*, (UK: Cambridge University Press, 2005), pp. 229–230; World Trade Organization, *A handbook on the WTO Dispute Settlement System* (UK: Cambridge University Press, 2004), pp. 60–62.

[100] Articles 10 and 11 of the 1992 OSPAR Convention.

[101] Article VI (1) (a) (b), (2), (3) and Annex III of the 1972 London Convention.

[102] Article X of the 1972 London Convention and Article 15 of the 1996 Protocol.

Article VI (4) states that it is the Contracting Party's obligation to report the information, criteria, measures and requirements it adopts to the IMO and other parties.[103] Article 9.4 of the 1996 Protocol echoes the ethos of the 1972 London Convention and states that these national governance instruments need to be submitted to the IMO or where appropriate, to other Contracting Parties on an annual or regular basis, which further strengthens the accountability of the Contracting Party. The reports will be reviewed by consultative meeting of the parties on an annual basis.[104] The content of the 1972 London Convention and its 1996 Protocol suggests that public authorities are responsible for controlling dumping behaviour. The monitoring process is by means of the IMO and other Contracting Parties. Externally, the Contracting Party is responsible to the IMO and other Contracting Parties. If the Contracting Parties fail in their responsibilities, the consultative meeting may develop or adopt procedures for determining exceptional and emergency situations.[105] Consultative meetings may also consider any additional action that may be required, if the aforementioned situation occurs.[106]

All Contracting Parties, under the MARPOL 73/78 must prohibit and take action against violations and accept certificates required by the regulations which are prepared by other parties as having the same validity as their own certificates.[107] A ship which is in the port or offshore terminal of a party may be subject to inspection, in order to verify the existence of a valid certificate, unless there are clear grounds for believing that the condition of the ship or its equipment does not substantially correspond with the particulars of that certificate.[108] If a certificate exists, the inspecting party shall, "ensure that the ship does not sail until it can proceed to sea without presenting an unreasonable threat of harm to the marine environment."[109] This interpretation suggests that the inspecting party should be responsible for taking any action against marine pollution or the threat thereof. If an inspection indicates a violation of the Convention, a report shall be forwarded to the Government of the State under whose authority the ship is operating.[110] The authority of the flag State shall promptly inform the inspecting party which has reported the alleged violation of the action taken, in addition to advising the IMCO.[111] By this means, the inspecting party and flag State are accountable to each other and to the IMCO.[112]

---

[103] Article VI (4) of the 1972 London Convention.
[104] Article XIV 4 of the 1972 London Convention.
[105] Article XIV 4 (e) of the 1972 London Convention.
[106] Article XIV 5 of the 1972 London Convention.
[107] Article 5.1 and 2 of the MARPOL 73/78.
[108] Article 5.2 of the MARPOL 73/78.
[109] Article 5.2 of the MARPOL 73/78.
[110] Article 6.2 of the MARPOL 73/78.
[111] Article 6.4 of the MARPOL 73/78.
[112] Article 4.3 and 6.4 of the MARPOL 73/78.

It is required that parties apply the MARPOL 73/78 to ships of non-parties, so as to ensure that no more favourable treatment is given to such ships.[113] There is also provision for the detection of violations and enforcement, such as in-port inspections, to verify whether ships have discharged harmful substances, reporting requirements on incidents involving harmful substances and the communication of information to the IMCO, plus technical co-operation.[114] The aforementioned treaty obligations form the accountabilities of the Contracting Parties.

The Contracting Parties must comply with the 1992 OSPAR Convention and the decisions and recommendations made by the Commission.[115] It is also the Contracting Parties' responsibility to report to the Commission at regular intervals on: "(a) the legal, regulatory, or other measures taken by them for the implementation of the provisions of the Convention and of decisions and recommendations adopted thereunder, including in particular, measures taken to prevent and punish conduct in contravention of those provisions; (b) the effectiveness of the measures referred to in subparagraph (a) of this Article; (c) problems encountered in the implementation of the provisions referred to in subparagraph (a) of this Article."[116] In return, the Commission shall assess the Contracting Parties' compliance with the Convention and the decisions and recommendations adopted thereunder.[117] If appropriate, the Commission may call for steps to bring about full compliance with the Convention and "promote the implementation of recommendations, including measures to assist a Contracting Party to carry out its obligations."[118] This conduct suggests that the Contracting Parties are responsible to the Commission, rather than to the general public. In practice, at the domestic level, each Contracting Party is responsible to its citizens, depending on its domestic law. The Convention uses the term "competent authorities",[119] to represent those domestic public authorities that are entitled to deal with marine pollution. In this way, the Convention does not distinguish jurisdictional boundaries between the various ocean governance authorities.

The International Court of Justice, in the *Nuclear Weapons* case, affirmed that the primary obligation of States is to ensure that their activities will not cause harm or potential harm to other States.[120] The substantive legal content of this customary rule is now evolving rapidly through the negotiation of specific treaties, some of which have been discussed in this section. As to the domestic level, some of the treaties do emphasise the importance of the competent authorities being

---

[113] Article 5.4 of the MARPOL 73/78.

[114] Article 6, 8, 11 and 17 of the MARPOL 73/78.

[115] Article 23 of the 1992 OSPAR Convention.

[116] Article 22 of the 1992 OSPAR Convention.

[117] Article 23.a of the 1992 OSPAR Convention.

[118] Article 23.b of the 1992 OSPAR Convention.

[119] Article 9 of the 1992 OSPAR Convention.

[120] *Nuclear Weapons* case ICJ Reports 1996, 266. See also *Nuclear Tests Examination Request New Zealand v France* ICJ Reports 1995, 288 at 306.

responsible for marine pollution but do not distinguish jurisdictional boundaries between the various ocean governance authorities. This, therefore, leaves room for States to decide their national ocean policy and to ensure that their legal systems provide for accountability to prevent and control marine pollution, for example, competent Ministers being answerable to the Parliament and the public. Following on from this, institutionally, States commit themselves in accordance with their policies, priorities and resources, to promote the institutional arrangements necessary to support the implementation of the programme areas regarding good ocean governance.[121]

2.3.6 Equity and Inclusiveness

The element of equity and inclusiveness aims to respect individual rights and interests. How does international treaty practice address the issue of equity and inclusiveness? The 1972 London Convention requires contracting parties to take all practicable steps to prevent the pollution of the sea.[122] In addition, the Convention and its 1996 Protocol, ask for the acknowledgement of characteristic regional features, when developing harmonising procedures.[123] Furthermore, the 1996 Protocol calls for Contracting Parties to take into account the special needs of developing countries and countries in transition to market economies.[124] This conduct should lead towards equitable and inclusive decision making but only at the international level, nevertheless, it gives an indication of the types of issues States ought also to take account of at the domestic level, when making decisions relating to ocean governance.

The 1992 Convention on Biological Diversity is aimed at "fair and equitable sharing of the benefits arising out of the utilization of genetic resources, [...], taking into account all rights over those resources and to technologies, and by appropriate funding."[125] To pursue the aforesaid objective, Contracting Parties should "respect, preserve and maintain knowledge, innovations and practices of indigenous and local communities embodying traditional lifestyles relevant for the conservation and sustainable use of biological diversity and promote their wider application with the approval and involvement of the holders of such knowledge, innovations and practices and encourage the equitable sharing of the benefits arising from the utilization of such knowledge, innovations and practices."[126] The 1995 United Nations Fish Stocks Agreement emphasises the need to ensure access

---

[121] Paragraph 116 of the Agenda 21.

[122] Article I of the 1972 London Convention.

[123] Article VIII of the 1972 London Convention and Article 12 of the 1996 Protocol.

[124] Article 13.1.5 of the 1996 Protocol.

[125] Article 1 of the 1992 Convention on Biological Diversity.

[126] Article 8 (j) of the 1992 Convention on Biological Diversity.

to fisheries by small scale and artisanal fishermen and female fish processors, as well as indigenous people in developing States, particularly in small, island developing States.[127]

The 1996 Protocol to the Convention on the Prevention of Marine Pollution by Dumping of Wastes and other Matter[128] represents a major change of approach to the question of how to regulate the use of the sea as a depository for waste materials and it is much more restrictive than the Convention. It encourages States to adopt "regional and national instruments which aim to protect the marine environment and which take account of specific circumstances and needs of those regions and States",[129] more importantly, "having due regard to the public interest."[130] To this end, it is clear that the 1996 Protocol requires taking decisions concerning all the varying circumstances.

The notion of equality is accepted by the States, in the sense of equality of legal personality and capacity. This notion was illustrated by the 1998 Aarhus Convention, as it states that it is necessary to encourage, "the possibility to participate in decision making and have access to justice in environmental matters without discrimination as to citizenship, nationality or domicile and, in the case of a legal person, without discrimination as to where it has its registered seat or an effective centre of its activities."[131]

To sum up, it can be seen that these international legal instruments providing an obligation to respect individual rights and interests. The utilisation of renewable resources[132] should be based on the principle of fairness and equality. Attention should be given to those people in the minority, such as indigenous people or female fish processors. The equality of legal personality and capacity will ensure the possibility of access to environmental justice, without discrimination. Yet again, these obligations provide an indication of the types of issues one would anticipate being taken into account in domestic decision making.

2.3.7 Responsiveness

Responsiveness emphasises the need for the public authorities to make decisions within a specific time scale. It is also important to provide a legal system/mechanism which can respond to the current needs of the general public/environment.

---

[127] Article 24.2 (b) of the 1995 United Nations Fish Stocks Agreement.

[128] The 1996 Protocol to the Convention on the Prevention of Marine Pollution by Dumping of Wastes and Other Matter (London), 7th November 1996, came into force on 24th March 2006, 36 ILM 1.

[129] Preamble of the 1996 Protocol.

[130] Article 3.2 of the 1996 Protocol.

[131] Article 3 (9) of the 1998 Aarhus Convention.

[132] This part of the research only considers renewable resources as, exhaustible resources would be subject to further research.

The following section will provide examples relative to responsiveness, from international treaty practice.

UNCLOS imposes an obligation on States, from time to time, to re-examine rules, standards and recommended practices and procedures, as regards the control and prevention of marine pollution.[133] This mechanism allows States to meet the requirement to be able to adapt to new issues, as they arise.

Under the 1992 Convention on Biological Diversity, Contracting Parties shall, "promote national arrangements for emergency responses to activities or events"[134] and also, "introduce appropriate arrangements to ensure that the environmental consequences of its programmes and policies that are likely to have significant adverse impacts on biological diversity, are duly taken into account."[135]

Under the 1995 United Nations Fish Stocks Agreement, States shall ensure that their decision making procedures facilitate the adoption of conservation and management measures, in a timely and effective manner.[136] There follows, some examples of how States should respond to the changing needs of the environment.

The 1992 OSPAR Convention requires the contracting parties to produce relevant information, "in response to any reasonable request, without that person's having to prove an interest, without unreasonable charges, as soon as possible and at the latest within two months."[137] This provision requires the Contracting Parties to treat all applications to dump on an equal basis, without discrimination and also requires States to make decisions within a specific time scale. In addition, the Contracting Parties are required to adopt the best available techniques and best environmental practice,[138] by which to ensure that their decision making system is responsive to the needs of individuals or the environment. To this end, the Convention not only illustrates a legal duty on public authorities to make decisions within a limited timeframe but also a duty to respond to the current needs of the general public and environment. This illustrates a good example of responsiveness in good ocean governance.

The 1998 Aarhus Convention imposes a further duty on public authorities, in that the latter shall make environmental information, "available as soon as possible and at the latest within one month after the request has been submitted, unless the volume and the complexity of the information justify an extension of this period up to two months after the request. The applicant shall be informed of any extension and of the reasons justifying it."[139] In the event of refusing a request, it is also the public authorities' duty to ensure that, "A refusal of a request shall be in

---

[133]   Article 207.4; 208.5 and 210.4 of UNCLOS.

[134]   Article 14 (e) of the 1992 Convention on Biological Diversity.

[135]   Article 14 (a) of the 1992 Convention on Biological Diversity.

[136]   Article 10 (j) of the 1995 United Nations Fish Stocks Agreement.

[137]   Article 9 (1) of the 1992 OSPAR Convention.

[138]   Article 2.2 of the 1992 OSPAR Convention.

[139]   Article 4 (2) of the 1998 Aarhus Convention.

writing if the request was in writing or the applicant so requests. A refusal shall state the reasons for the refusal and give information on access to the review procedure." In addition, "the refusal shall be made as soon as possible and at the latest within one month, unless the complexity of the information justifies an extension of this period up to two months after the request. The applicant shall be informed of any extension and of the reasons justifying it."[140] This offers a good example of responsiveness, in terms of decision making.

A number of treaties have, in different ways, emphasised the importance of the element of responsiveness in good governance. Following on from the above discussion, it is evident that international law requires States to make decisions within a specific time scale, depending on the specific circumstances involved. In addition, international law also requires States to take all practicable measures to ensure that the current needs of the general public and environment can be met. The latter is reflected in Chapter 17 of Agenda 21 which indicates that States should, "design and implement rational response strategies to address the environmental, social and economic impacts of climate change and sea level rise, and prepare appropriate contingency plans."[141] One would anticipate that such provisions should be mirrored in domestic provisions.

2.3.8 Coherence

Coherent decision making emphasises the requirement that decisions are consistent and make sense across time and institutions. To achieve this objective may require an institution to coordinate various public authorities, in order to facilitate a comprehensive ocean governance policy. The next section will provide examples addressing the element of coherence from international treaty practice.

Under the 1992 Convention on Biological Diversity, each Contracting Party shall "Integrate, as far as possible and as appropriate, the conservation and sustainable use of biological diversity into relevant sectoral or cross-sectoral plans, programmes and policies"[142] and introduce this concept into the national decision making processes.[143]

The Commission to the 1992 OSPAR Convention provides a forum for all the Contracting Parties and for a review of, "the condition of the maritime area, the effectiveness of the measures being adopted, the priorities and the need for any additional or different measures."[144]

---

[140] Article 4 (7) of the 1998 Aarhus Convention.
[141] Paragraph 17.128 (g) and see also 17.100 (c) of the Agenda 21.
[142] Article 6 (b) of the 1992 Convention on Biological Diversity.
[143] Article 10 (a) of the 1992 Convention on Biological Diversity.
[144] Article 10.2.b of the 1996 OSPAR Convention.

Chapter 17 of Agenda 21 states that States should take effective action consistent with international law, so as to monitor and control the activities to ensure compliance with applicable conservation and management rules, including full, detailed, accurate and timely reporting, in order to facilitate good governance.[145]

Although international treaty practice emphasises the need for integration of national ocean policy, they do not indicate how this objective might be achieved, save to suggest the need to monitor activities, in order to ensure compliance with international obligations. One option to ensure coherence might be to establish a cross-sectoral body, in order to oversee ocean governance issues, although, currently, there is no clear example of this to draw from international law.

2.4 Conclusion

The various treaties and soft law instruments referred to above provide an indication of what good ocean governance may look like in practice. It is clear that the implementation of good ocean governance at the national level may vary slightly but the lessons that can be learnt include the fact that law should be published in an appropriate way and should be equally and fairly administered and effectively enforced. The decision makers are, therefore, bound to follow the rule of law when making decisions.

There is a growing awareness as regards the desirability of including public participation in the decision making processes. The need for public participation is not, however, universally acknowledged. Consultation is largely limited to the Contracting Parties, organisations and experts but does not include the general public. It is anticipated that participation in terms of decision making at the national level is likely to be limited to the relevant stakeholders and experts.

Attention is devoted to transparent decision making, which may require a considerable amount of environmental information to be released to the public. To release environmental information to the general public is the central concern of the 1998 Aarhus Convention, whereas, other treaties are more concerned with the making of information available to other Contracting Parties and Organisations. The key lesson from international law seems to be that information must be made available to all those who may have an interest in it, be that the general public or specific stakeholders. It is observed that there is a growing movement toward making environmental information available to the general public. This statement, however, needs to be further scrutinised at national level.

Individual rights and interests are emphasised in the international legal order and the same practice ought to be evident at national level. International treaty practice emphasises the need to make decisions within a specific time scale, depending on specific circumstances. In addition, international law requires States

---

[145] Paragraph 17.51, 68,108,116, of the Agenda 21.

to take all practicable measures to ensure that the current needs of the general public and environment can be met. It is, therefore, not unreasonable to suggest that decision making at the national level should also be on a time related basis and should reflect the needs of the general public and the environment.

International trade law clearly perceives consensus as meaning no member formally objecting to the proposed decision at a meeting, when a decision is taken. To seek common agreement in the decision making processes is essential and therefore, needs to be further addressed at the domestic decision making level. Under customary international law, States are accountable for their surrounding marine environment and are entitled to adopt all appropriate measures to maintain environmental quality and control marine pollution. International law does not, however, address the issue whereby public authorities should be answerable to Parliament and the general public.

Finally, according to international treaty practice, States should integrate national ocean policy, in order to coordinate different sectors of ocean governance programmes. The result of the aforesaid should be not only more effective governance of the oceans at national level but also a uniform and consistent national position at regional and global levels. The consequence of this would foster better cooperation among States, as well as between international organisations, when addressing oceans issues, potentially leading to more integrated and more effective ocean governance at the global level.

Chapter 3
Ocean Governance: It is Time to Change

Abstract Traditionally, the public authorities with their sectoral arrangements and ways of conducting business, constrain efforts at suitable evaluation regarding appropriate ocean governance actions. It is for these reasons that 'joined-up' efforts began to emerge. Such efforts are likely to cut across spatial boundaries in the context of marine ecosystems, reflecting an evolving realisation of how the marine environment operates, which will enable the public authorities to adapt appropriate human behaviour and governance systems. Clearly, it is insufficient to concentrate only on particular ocean uses in isolation. It is, instead, necessary that ocean governance be pursued in the context of the holistic use of ocean space with an understanding of use interactions and the cumulative effects of human activities on the marine environment. This in turn indicates that certain degree of change is needed. This chapter suggests that there is a need for a change in term of attitude, legal authority and institutional structure with references to the United States, Canada and Australia.

Keywords Ocean governance · Changing the attitude · Changing legal authority · Changing institutional structure

3.1 Introduction

Most efforts directed towards ocean governance concentrate on the *what* questions e.g. what elements contribute to ocean governance; what element of ocean governance is missing in State practice. In order to make real progress, attention should also focus on the *how* problems which will consider to what extent ocean governance policy should be delivered. The answers of these questions or problems require a greater effort to establish an appropriate basis for ocean governance. The primary issue is that the concept of good ocean governance should be incorporated into the public authorities' decision-making processes.[1] This action will require the public authorities to follow open, clear and stable rule of law; open their decision-making process in order to allow the broadest possible participation; pro-actively release environmental information to encourage consensus based decision-making; the public authorities are answerable to the Parliament and the general public; decisions take due account of all aspects of interests and within a specific limit of timeframe and overall respond to the current needs of the social, economic and the marine environment.

Secondly, the aforementioned obligations should be followed by environmental law in order to impose legal restriction. The Stockholm Declaration in 1972[2] is considered to be a benchmark for the start of the current environmental move-ment.[3] It not only energised governments but also helped to encourage public interest in public authorities devoted to various types of environmental causes.[4] Maritime sustainability is deemed to be essential to the future wellbeing of the world. Aspects of this can be seen in the 1992 Rio Declaration, Chapter 17 of Agenda 21,[5] the 2002 Johannesburg Declaration,[6] the 1982 Convention on the Law of the Sea (UNCLOS),[7] the 1972 Convention on the Preservation of Marine Pollution by Dumping of Waste and Other Matter,[8] the 1992 Convention for the Protection of Marine Environment of the North-East Atlantic[9] as well as the International Convention for the Prevention of Pollution from Ships 1973 together with its modified convention is known as the MARPOL 73/78. Most of these legal instruments provide partial answers to the subject they deal with but an overall sense of direction is still missing.

In addition, it is ideal to wish that individuals and enterprises should feel responsible for carrying out their maritime activities in a sustainable manner, but *who* should hold them accountable? Who can respond to marine issues quickly and efficiently and can integrate across various barriers—the reasonable answer is the

---

[1] Yen-Chiang Chang, "Good Ocean Governance", in *Ocean Yearbook* 23, (Canada: Dalhousie Law School, 2009), pp. 89–118.

[2] The United Nations Conference on the Human Environment, met at Stockholm from 5 to 16 June 1972, full content available from: http://www.unep.org/Documents.multilingual/Default.asp?DocumentID=97&ArticleID=1503 last visited date: 07/05/2009.

[3] Robert Friedheim, "Designing the Ocean Policy Future: An Essay on How I Am Going To Do That", *Ocean Development & International Law* 31 (2000):185–186.

[4] Ibid.

[5] United Nations, *Report of World Summit on Sustainable Development*, Johannesburg, South Africa, 26 August–4 September 2002, ISBN 92-1-104521-5, p. 8.

[6] Adopted at the 17th plenary meeting of the World Summit on Sustainable Development, on 4 September 2002, see Chap. VIII of the Summit Report.

[7] Untied Nations Convention on the Law of the Sea, Montego Bay, 10 December 1982, came into force 16 November 1994, 21 ILM 1245 (1982).

[8] Convention on the Preservation of Marine Pollution by Dumping of Waste and Other Matter, (London, Mexico City, Moscow, Washington DC) 29 December 1972, came into force 30 August 1975; 1046 UNTS 120.

[9] Paris, 22 September 1992, came into force 25 March 1998, 23 LOSB 32.

government. It is perceived that in China the public authorities concerned with marine natural resources and ocean/coastal-related activities have been given sectoral jurisdiction and responsibilities. The main criticism as regards the ocean governance was the lack of coherent, coordinated mechanisms operated by the executive branch. No one was solely in charge, in term of both responsibility and authority. That is to say, the public authorities in China oversaw a particular type of activity such as marine pollution control, fishing, navigation and mining which indicated a narrow focus. The problems had been identified but there seemed to be no effort within the government to organise an appropriate solution.

Where marine resources are immense and interactions among parts of the natural ecosystem or different interests groups are nonexistent, such an approach to ocean governance might be tenable.[10] In reality, it is clear that sectoral approaches to the marine environment and its resources are dysfunctional because of mutual interference among different interests groups.[11] It is for these reasons that 'joined-up' efforts began to emerge. Such efforts are likely to cut across spatial boundaries in the context of marine ecosystems, reflecting an evolving realisation of how the marine environment operates, which will enable the public authorities to adapt appropriate human behaviour and governance systems.

Clearly, it is insufficient to concentrate only on particular ocean uses in isolation. It is, instead, necessary that ocean governance be pursued in the context of the holistic use of ocean space with an understanding of use interactions and the cumulative effects of human activities on the marine environment. Traditionally, the public authorities with their sectoral arrangements and ways of conducting business, constrain efforts at suitable evaluation regarding appropriate ocean governance actions. The introduction of this chapter have suggested *what* should be done but still flounder over the questions of *who* should do it and *how*. The answers to *who* and *how* questions will be detailed later in this chapter. In sum, this chapter suggests that to achieve good ocean governance, there is a need for a change in term of attitude, legal authority and institutional structure with references to the United States, Canada and Australia.[12]

---

[10] Yen-Chiang Chang, "A Study of Marine Resources Governance", *Taiwan Ocean Law Review* 7 (2008):111–132, Lawrence Juda, "Changing National Approaches to Ocean Governance: The United States, Canada, and Australia", *Ocean Development & International Law* 34 (2003):162.

[11] The same opinion please see K. C. Roy and C. A. Tisdell, "Good Governance in Sustainable Development: The Impact of Institutions", p. 13, in the book edited by Robin Ghosh, Rony Gabbay, Abu Siddique, *Good Governance Issues and Sustainable Development—The Indian Ocean Region* (Atlatic Publishers and Distributors, 1999).

[12] Lawrence Juda, supra note 10, p. 162.

3.2 Changing the Attitude

To 'integrate' means to combine parts into a whole, to complete something that is
imperfect or incomplete by adding parts.[13] Integrated marine policy, thus, means a
marine policy designed to overcome the fragmentation inherent both in single-
sector governance approaches (e.g. fishing, oil and gas extraction and marine
transport) and in the splits in jurisdiction among different levels of government.[14]
To qualify as an integrated marine policy, any proposal need to meet three
requirements, namely be comprehensive in scope, aggregate in assessment and
consistent in implementation. These requirements refer to three consecutive stages
of the decision-making process: comprehensive in scope to the input stage;
aggregate in assessment to the processing of inputs and consistent in implementation to the output stage.[15]

3.2.1 Comprehensive in Scope

Scope can be measured in four different ways, videlicet: time, space, actors and
issues. Evaluating policy alternatives not only on their short-term merits but also
on the basis by which consequences may accrue in the distant future. An integrated
marine policy generally refers to long-term ocean governance activity. For
example, to govern fish stocks may require adopting exploitation limits such as a
maximum sustainable yield.[16]

From a spatial perspective, a common marine environmental policy for the
entire North Sea would be more 'integrated' than a set of national ocean policy
coving only a national maritime zone.[17] In practice, the separation of ocean space
into national jurisdiction does not always correspond to distinct marine ecosystems
which may in turn lessen the impact of the overall ocean governance activities. For
example, in 2003, the European Commission adopted measures under Council
Regulation 2287/2003[18] to impose a spatial management plan for threatened
species based on a set of closed or semi-closed areas. The Regulation limits fishing
effort in these areas, with specific control and monitoring rules to ensure implementation. The stocks covered by the plan include cod in the Kattegat, the North

---

[13] A. S. Hornby, *Oxford Advanced Learner's Dictionary of Current English*, (Taipei: Dong-Hua
Publisher 1989), p. 603.

[14] Biliana Cicin-Sain and Robert W. Knecht, *Integrated Coastal and Ocean Management—
Concepts and Practices* (Island Press, 1998), p. 1.

[15] Arild Underdal, "Integrated Marine Policy—What? Why? How?", *Marine Policy* 4 (3) July
1980:159.

[16] Ibid, p 160.

[17] Ibid.

[18] Council Regulation COM 2287/2003. Official Journal L344/1. 31.12.2003.

Sea including the Skagerrak and the Eastern Channel, the west of Scotland and the Irish Sea.[19]

Regarding the actors in the ocean governance area, policy scope can be determined as the proportion of actors within a given ocean governance activity that is included in the 'reference group'. This is to say, the group from whose perspective policy options are being evaluated. Regarding the issue dimension, a State governing its marine environmental protection, fisheries, offshore gas and oil extraction and marine transport under a common policy framework can be said to have a more comprehensive marine policy than a State treating each of these activities as a distinct area.[20] As previous discussion suggests that in order to move towards more appropriate ocean governance policy, the various parties should be given the necessary opportunity to enable them to present their cases effectively.[21] This would mean that judges and decision makers should consider all relevant options. Therefore, the issue that needs to be considered is to ensure that individual rights and interests are properly respected.

3.2.2 Aggregate in Assessment

The aggregation of decision-making refers to the extent to which policy alternatives are assessed from an 'overall' perspective, rather than from the view of each actor or sector. An overall perspective is often very difficult to achieve, in particular at the international level. International ocean governance seems, in most cases, to suffer more from aggregation failures than from a lack of comprehensiveness.[22] The aforesaid situation is no better at the domestic level, since the expansion of government often takes place through so called 'cell formation'.[23] It is, therefore, important to note that the trade off involved which would require sacrifices by some actors for the common good, is a situation where some form of compensation may be required. Another implication of this is that the aggregation of policy making is not a purely technical exercise, it implies weighting interests and setting priorities.[24]

At the moment, the ocean governance system in China is organised in a way of fragmentation, therefore, control tends to rely on self-discipline. Each department works towards its own objectives in its own way. The culture of monolithic, inward-looking decision-making may have a negative influence on producing an integrated national ocean policy.

---

[19] Ibid.

[20] Arild Underdal, supra note 15, p. 160.

[21] Andrew Le Sueur and Maurice Sunkin, *Public Law* (London: Longman, 1997), p. 583.

[22] Arild Underdal, supra note 15, p. 161.

[23] Ibid.

[24] Ibid.

3.2.3 Consistent in Implementation

Two dimensions need to be considered when discussing marine policy consistency. From a vertical aspect, a consistent marine policy can be defined as one in which specific implementation measures conform to more general guidelines and objectives. As regards the horizontal dimension, consistency can be understood as a requirement that only one policy is being pursued at a time by all relevant public authorities.[25] Consistency often implies equal treatment of equal cases. This means applying the same substantive principles and criteria to all cases in a certain category. However, the requirement of consistency should not be deemed to imply that policy should not change as a response to changing circumstances or new information.

3.2.4 Summary

An appropriate integrated marine policy can be defined as one in which all significant consequences of decisions should be considered as decision premises, where policy options are evaluated on the basis of an aggregate measure of utility and where the different policy elements are consistent with each other. To achieve an integrated marine policy, two options should be considered. Firstly, to establish precise integrating objectives and guidelines and to ensure their application. The more vague and general the principles, the less guidance will be provided. The more clear and specific the guidelines, the more effort and energy will go into their formulation.[26] Secondly, is the indirect strategy, this being 'intellectual' and 'institutional'. The intellectual strategy seeks integrated marine policy through initiating research, training and socialisation aimed at developing a more comprehensive perspective of decision-making.

3.3 Changing Legal Authority

This section will start with the question, 'What is legal authority?' The general public continually makes rules for particular circumstances. A freely-negotiated commercial contract may bind contracting parties to behave in a particular way. By becoming members of a legal association or a sports club, there may be a requirement to comply with a set of rules. In certain circumstances, these forms of law are used by the courts to enforce the agreements. Law is, therefore, a way of regulating behaviour, of deciding what can be done and what cannot be done.[27] To put it in a simple way, law is 'the rules of the game'.

---

[25] Arild Underdal, supra note 15, p. 161.

[26] Ibid, p. 166.

[27] A. Bradney, V. Fisher, J. Masson, A. Neal and D. Newell, *How to Study Law* (London: Sweet & Maxwell, 1991), p. 3.

The rules of the game at seas were introduced by UNCLOS to ensure that the oceans are well governed and protected. To achieve the objective stated by UNCLOS, an integrated marine policy is considered to be a solution to marine pollution control and to resolve conflicts between different interests. Within a coastal State's jurisdiction, a single 'Marine Act' is likely to create an integrated approach to the governance of human activities, engaging stakeholders and realising the economic potential.[28] At the same time, a Marine Act would provide the legal basis for an integrated marine policy. As marine issues are traditionally governed on a sectoral basis, in some coastal States, the law in these States is also on a sectoral basis e.g. Taiwan and the United Kingdom. There is, therefore, a need to examine other major coastal States such as the United States, Canada and Australia, in order to determine whether they operate on a sectoral basis or whether they have a single Marine Act. In the event of the latter, the efficiency of the single act will be examined.

3.3.1 The United States

In the early stages of the United States' coastal and ocean legislation, the individual sector approach is evident in the Clean Water Act,[29] the Marine Mammal Protection Act,[30] the Endangered Species Act,[31] the Ocean Dumping Act,[32] Resource Conservation and Recovery Act[33] and the Fisheries Conservation and Management Act.[34] As noted by Cicin-Sain and Knecht, the individual sector-by-sector approach represented by such legislation failed to establish priorities among the competing uses of coastal and ocean space.[35] This situation leads to the view that such legislation makes a limited contribution to resolving the conflict arising from the use of coastal and ocean resources.

One important exception to the sectoral basis approach in the United States is the Coastal Zone Management Act,[36] which pursues a spatial but not a sectoral approach to the governance of coastal zones. Important incentives in the form of federal funding and federal consistency are to encourage states to adopt

---

[28] Malcolm MarGarvin, *A Marine Act? For the UK*, Modus Vivendi for WWF UK, December 2000, pp. 3–5.

[29] The Clean Water Act 1972, 33 U.S.C. § 1251 et seq.

[30] The Marine Mammal Protection Act 1972, 16 U.S.C. 1361.

[31] The Endangered Species Act 1973, Public Law 93–205.

[32] The Ocean Dumping Act 1972, 33 U.S.C. § 1401 et seq.

[33] The Resource Conservation and Recovery Act 1976, 42 U.S.C. s/s 6901 et seq.

[34] The Fisheries Conservation and Management Act 1976, Public Law 94-265, approved April 13, 1976; 16 U.S.C. 1801–1882.

[35] Biliana Cicin-Sain and Robert W. Knecht, supra note 14, pp. 160–166.

[36] The Coastal Zone Management Act 1972, Public Law 92–583.

comprehensive coastal zone management plans. The expected outcome of this Act would be to meet broad guidelines and address issues of conflict of use and environmental damage in coastal zones. In the respect of the comprehensive spatial based approach, the Coastal Zone Management Act may be seen as a major 'precursor' to the development of ecosystem-based approach.[37]

Another piece of legislation that contributes to a broader evaluation of ocean and coastal activities is the National Environmental Policy Act,[38] which requires federal agencies to integrate environmental values into their decision making processes by considering the environmental impact of their proposed actions and considering reasonable alternatives to those actions.[39] To meet this requirement, federal agencies need to prepare a detailed statement known as the Environmental Impact Statement. The Environmental Protection Agency then reviews and comments on the Environmental Impact Statement prepared. The Environmental Protection Agency is obliged to give access to the said Statements to the public.[40]

As a result of the above, the sectoral basis decision-making had been widened and a variety of requirements for consultation across federal public authorities and between the federal and state governments has been established.[41] Nonetheless, all of the above mentioned legislation provides a real but limited movement toward an integrated ocean policy. In the absence of a single comprehensive ocean policy, attempts are being made to provide for multi-use accommodation, which would cut across various public authorities.

Later, in 2000, the Oceans Act was introduced to establish a new Commission on Ocean Policy to make recommendations for "coordinated and comprehensive national ocean policy".[42] To this end, the following objectives are pursued: protection of life and property; stewardship of ocean and coastal resources; protection of marine environment and prevention of marine pollution; enhancement of maritime commerce; expansion of human knowledge of the marine environment; investments in technologies to promote energy and food security; close cooperation among government agencies and the United States leadership in ocean and coastal activities.[43]

The Oceans Act forced the Commission to deliver a report to the Congress and President within 18 months of the Commission's establishment.[44] The said report reviewed and made recommendations with respect to:

---

[37] Lawrence Juda, supra note 10, p. 166–167.

[38] The National Environmental Policy Act 1969, Public Law 91–190.

[39] See U.S. Environmental Protection Agency official website: http://www.epa.gov/compliance/nepa/last visited date: 07/05/2000.

[40] Ibid. See also Lawrence Juda, supra note 10, p.167.

[41] Lawrence Juda, supra note 10, p. 167.

[42] U.S. Commission on Ocean Policy official website: http://www.oceancommission.gov/documents/oceanact.html last visited date: 07/05/2009.

[43] Ibid.

[44] Section 3 (f) (1) of the Oceans Act 2000.

"A. An assessment of facilities (people, vessels, computers, satellites);

B. A review of federal activities;

C. A review of the cumulative effect of federal laws;

D. A review of the supply and demand for ocean and coastal resources;

E. A review of the relationships between federal, state, and local governments, and the private sector;

F. A review of the opportunities for the investment in new products and technologies recommendations for modifications to federal laws and/or the structure of federal agencies;

G. A review of the effectiveness of existing federal interagency policy coordination."[45]

Furthermore, the Oceans Act insists that the Commission opens *ALL* meetings to the public and therefore, interested persons shall be permitted to appear at open meetings and present oral or written statements on the subject matter of the meeting.[46] The minutes and records of all meetings and other documents that were made available to or prepared for the Commission shall be available for public inspection.[47] Prior to delivery of the final report, the draft report was to be made available for review by the public and the governors of coastal states.[48] Within 90 days of receipt of the Commission report, the President is required to provide Congress with a statement of proposals to implement or respond to "the Commission's recommendations for a coordinated, comprehensive, and long-range national policy for the responsible use and stewardship of ocean and coastal resources for the benefit of the United States."[49] Study of the Oceans Act suggests that it is the Congress's intention to obtain recommendations in order to develop an integrated ocean policy. The Act is also seen to be a catalytst to facilitate a better approach to ocean governance.

3.3.2 Canada

Similar to the United States, the early stages of the Canadian federal legislation pertaining to ocean governance were developed in a piecemeal fashion, with a combination of sectoral and functional bases.[50] This is evidenced by the Fisheries Act,[51] the Oil and Gas Production and Conservation Act,[52] the Canada Water Act,[53] the Territorial Sea and Fishing Zones Act,[54] the Coastal Fisheries Protection

---

[45] Section 3 (f) (2) of the Oceans Act 2000.

[46] Section 3 (e) (1) of the Oceans Act 2000.

[47] Section 3 (e) (2) of the Oceans Act 2000.

[48] Section 3 (g) (1) (2) of the Oceans Act 2000.

[49] Section 4 (a) of the Oceans Act 2000.

[50] Edited by David VanderZwaag, *Canadian Ocean Law and Policy* (Toronto: Butterworths, 1992), p. 329.

[51] The Fisheries Act, R. S. C. 1985, c.F-14, s. 36.

[52] The Oil and Gas Production and Conservation Act, R. S. C. 1985, c. O-7.

[53] The Canada Water Act, R. S. C. 1985, c. C-11.

[54] The Territorial Sea and Fishing Zones Act, R. S. C. 1985, c. T-8.

Act,[55] the Pest Control Products Act,[56] the Fishing and Recreational Harbours Act,[57] the Navigable Waters Protection Act,[58] the Canadian Environmental Protection Act,[59] the Canada Shipping Act,[60] the Arctic Waters Pollution Prevention Act,[61] the Canada Petroleum Resources Act,[62] the Public Harbours and Port Facilities Act,[63] the Canada Ports Corporation Act[64] and the National Parks Act.[65]

Among the aforementioned legislation, few of the statutes focus entirely on the marine environment but some of them only apply in a very peripheral manner e.g. the Pest Control Products Act.[66] Part of the legislation is designed mainly to provide a framework in which a particular sectoral activity can be operated in a safe manner e.g. the Canada Shipping Act, the Public Harbours and Port Facilities Act and the Oil and Gas Production and Conservation Act. Some of the legislation is introduced to constrain jurisdiction over one or more activities e.g. the Arctic Waters Pollution Prevention Act and the Territorial Sea and Fishing Zones Act. It has been argued that the lack of coherence in marine pollution control due to the mix of regulations and guidelines, has placed emphasis on control rather than prevention and the division of the public authorities between different interests.[67]

At the present stage of legislative development, the Canadian Environmental Protection Act[68] and Oceans Act[69] are considered as providing the integration and consistency necessary to reduce the fragmentation that currently exists. Under the Canadian Environmental Protection Act, environmental standards, ecosystem objectives and environmental quality guidelines and codes of practice can be developed and will apply to all public authorities.[70] The Act is "based on an ecologically efficient use of natural, social and economic resources and acknowledges the need to integrate environmental, economic and social factors in the making of all decisions by government and private entities.".[71]

---

[55] The Coastal Fisheries Protection Act, R. S. C. 1985, c. C-33.

[56] The Pest Control Products Act, R. S. C. 1985, c. P-9.

[57] The Fishing and Recreational Harbours Act, R. S. C. 1985, c. F-24.

[58] The Navigable Waters Protection Act, R. S. C. 1985, c. N-22.

[59] The Canadian Environmental Protection Act, R. S. C. 1985 (4th Supp.), c.16.

[60] The Canada Shipping Act, R. S. C. 1985, c. S-9.

[61] The Arctic Waters Pollution Prevention Act, R. S. C. 1985, c. A-12.

[62] The Canada Petroleum Resources Act, R. S. C. 1985 (2nd Supp.), 36.

[63] The Public Harbours and Port Facilities Act, R. S. C. 1985, c. P-29.

[64] The Canada Ports Corporation Act, R. S. C. 1985, c. C-9.

[65] The National Parks Act, R. S. C. 1985, c. N-14.

[66] Edited by David VanderZwaag, supra note 50, p. 329.

[67] P. Muldoon and M. Valiante, *Toxic Water Pollution in Canada: Regulatory Principles for Reduction and Elimination* (Calgary: The Canadian Institute for Resources Law, 1989), p. 10–11.

[68] The Canadian Environmental Protection Act, 1999, c. 33.

[69] The Oceans Act, 1996, c. 31.

[70] Preamble of the Canadian Environmental Protection Act.

[71] Ibid.

The Oceans Act articulates a new approach to ocean policy in which it indicates a desire to move beyond sectoral approaches to favour developing an integrated ocean policy, with an emphasis on the ecosystem-based and precautionary approaches, so as to ensure marine biodiversity and productivity.[72] The concepts of sustainable development, integrated governance and precautionary approaches are the basic principles of the Oceans Act.[73] With the aforesaid concepts in mind, the Act calls for the Minister of Fisheries and Oceans, "in collaboration with other ministers, boards and agencies of the Government of Canada, with provincial and territorial governments and with affected aboriginal organizations, coastal communities and other persons and bodies, including those bodies established under land claims agreements, shall lead and facilitate the development and implementation of a national strategy for the management of estuarine, coastal and marine ecosystems in waters that form part of Canada or in which Canada has sovereign rights under international law."[74]

In exercising the powers and performing the duties and functions mentioned in the Oceans Act, the Minister of Fisheries and Oceans shall "develop and implement policies and programs with respect to matters assigned by law to the Minister"[75] and shall also "coordinate with other ministers, boards and agencies of the Government of Canada the implementation of policies and programs of the Government with respect to all activities or measures in or affecting coastal waters and marine waters."[76] The Minister, for the purposes of the implementation of integrated governance plans, may at his/her discretion consult "another person or body or with another minister, board or agency of the Government of Canada, and taking into consideration the views of other ministers, boards and agencies of the Government of Canada, provincial and territorial governments and affected aboriginal organizations, coastal communities and other persons and bodies, including those bodies established under land claims agreements."[77]

The dimensions of such a strategy cut across political and bureaucratic jurisdictions in favour of the ecosystem-based approach. A basic premise of the Oceans Act is that an ocean governance strategy will require a collaborative and inclusive effort among stakeholders, both inside and outside of government.[78] The Minister of Fisheries and Oceans has been given the supreme authority, under the Oceans Act, to develop and implement the integrated ocean policy in which ocean governance jurisdiction is spread over a number of public authorities.

---

[72] Preamble of the Oceans Act.

[73] Section 30 of the Oceans Act.

[74] Section 29 of the Oceans Act.

[75] Section 32 (a) of the Oceans Act.

[76] Section 32 (b) of the Oceans Act.

[77] Section 32 (c) (d) of the Oceans Act.

[78] Section 31–33 of the Oceans Act.

3.3.3   Australia

In Australia, the Offshore Constitutional Settlement[79] is known as a milestone in intergovernmental negotiations and provides the framework for ocean governance.[80] Under the Offshore Constitutional Settlement, the states and territories of Australia have the power to govern coastal and ocean activities from the low-water mark to the three nautical miles. Meanwhile, the central or Commonwealth Government has the authority and responsibility beyond the three nautical miles, regarding ocean governance activities.[81] The development and implementation of the Offshore Constitutional Settlement was unique in that it addressed marine issues separately within its 'agreed arrangements'.[82] These arrangements including a legislative package, an offshore petroleum package, an offshore fisheries package, a Great Barrier Reef package and new ancillary arrangements.[83] It is on this basis that the Offshore Constitutional Settlement makes up the legislative framework for the development and implementation of Australia's Oceans Policy. The aforesaid Settlement also secures the Commonwealth Government's position as a major decision maker in ocean governance, which has dominated the ocean policy community.

Environment Australia reviewed the existing Commonwealth legislative framework and concluded that ocean governance legislation was fragmentary and complex, actively hindering the development of an integrated approach to resource allocation, management and conservation.[84] The current problems of Australian's ocean governance legislation have been identified as: "a general failure to address multiple use management; single-activity legislation that impedes integrated management; such multiple use legislation that exists is restricted to conservation; the absence of any legal framework for conflict resolution or avoidance; and 'recent' economic activities (tourism, recreation) that have no legislative basis."[85] It is also understood that "legislative change cannot be expected in the short term

---

[79] Marcus Haward, "The Offshore Constitutional Settlement", Marine Policy 13 (1989):334–348; Marcus Haward, "Developing an Australia Oceans Policy", in Elizabeth Mann Borgese, Aldo Chircop and Moira McConnell, Ocean Yearbook 15 (University of Chicago Press, 2001), p. 531–532.

[80] Joanna Vince, "The South East Regional Marine Plan—Implementing Australia's Oceans Policy", Marine Policy 30 (2006):421.

[81] Joanna Vince, supra note 80, p. 421; see also Lawrence Juda, supra note 10, p. 177.

[82] Ibid.

[83] Commonwealth of Australia, The Offshore Constitutional Settlement—A milestone in Cooperative Federalism (Canberra: AGPS, 1980), p. 1–2.

[84] Environment Australia, Multiple Use Management in the Australia Marine Environment—Principles, Definitions and Elements, Canberra, available from: http://www.environment.gov.au/marine/frameset/oceans/fs_ocean_issue_main.html website no longer exist, copy in file.

[85] Malcolm MacGarvin, supra note 28, p. 31–32.

[so] multiple use management will need to progress under the existing legislative framework while a better framework is developed".[86]

The proposed Marine Act is said to be one of the solutions to the above mentioned problems.[87] However, the enactment of a Marine Act is not currently being considered within the Australia's Oceans Policy process.[88] Furthermore, given the lack of a statutory base, the Australia's Oceans Policy remains a 'soft instrument' and does not seem to have the power to constrain the states of Australia. All of the federal government's agencies, indeed, are affected by the Australia's Oceans Policy. However, the Ministerial Council does not have the power to direct the work of public authorities outside of the Council membership. The Ministerial Council can, in fact, only seek cooperation.

3.3.4 Summary

As can be seen from the above discussion, the current practice of aforesaid States as regards ocean governance is moving toward a formally defined, legislated, integrated marine policy. An integrated marine policy has to be implemented through a marine agency or across the full range of integrated public authorities and on basis of appropriate scales of the marine ecosystem. Given legal status by means of a Marine Act not only creates an integrated ocean policy but also provides the legal basis for the aforesaid policy. The issue which needs to be considered is whether public authorities should be responsible for developing and implementing the integrated ocean policy and this issue will be discussed in the subsequent part.

3.4 Changing Institutional Structure

As pointed out by the Independent Commission on the World's Ocean, the fragmentation of ocean law and institutions at both national and international levels is a major obstacle to effective ocean governance. While recognising that there was no single, optimal model that all states should adopt, the Commission strongly emphasised the need for state to "establish at a high governmental level an appropriate policy and coordinating mechanism, to set and review national goals for ocean affairs."[89] The aforesaid body was deemed to be able to reconcile

---

[86] Environment Australia, *Multiple Use Management in the Australia Marine Environment— Principles, Definitions and Elements*, Canberra, available from: http://www.environment.gov.au/marine/frameset/oceans/fs_ocean_issue_main.html website no longer exist, copy in file.

[87] Smyth C, Haward M, Davey K, Grady M, *Oceans Eleven* (Victoria: Australian Conservation Foundation, 2003), p. 12–13.

[88] Joanna Vince, supra note 80, p. 425.

[89] Independent World Commission on the Ocean, *The Ocean: Our Future* (Cambridge: Cambridge University Press, 1998), p. 152.

policies and programmes in order to promote an integrated approach to ocean governance. At present, efforts are being made to develop new policies and institutional arrangements for effective and efficient ocean governance and to establish integrated ocean policies to overcome the problems of sectoral based approaches to the ocean.[90] The task of this is proving to be difficult.

What is possible is an examination of the attempts being made in other coastal States such as the United States, Canada and Australia, in order to develop more holist and comprehensive ocean governance approaches. Integrated approaches to the ocean governance can be pursued through institutional change in government involving efforts towards coordination and centralisation of governmental functions regarding ocean governance.

3.4.1 The United States

In 1966, the Commission on Marine Science, Engineering and Resources was established to conduct "a comprehensive investigation and study of all aspects of marine science in order to recommend an overall plan for an adequate national oceanographic program that will meet the present and future national needs."[91] The Commission, chaired by Julius Stratton, was known as the Stratton Commission, issued an influential report, *Our Nation and the Sea*, in January 1969.[92] The Commission suggested that the legal responsibility for ocean activities was spread among a number of public authorities with overlapping jurisdiction, thus leading to conflict. Some important ocean-related programmes carried out by this ocean governance structure were seen to be of only marginal importance.[93] What had happened was that as a particular ocean governance problem arose, such as marine pollution control, fishing, navigation or offshore oil and gas development but it was addressed in isolation. Without reference to the marine environment and its uses taken as a whole, this was said to have a negative effect on long-term and rational ocean governance.[94] In order to solve this problem, the Commission suggested the creation of a new independent agency which would provide a central and integrated concentration on relevant ocean governance programmes.[95]

The National Oceanic and Atmospheric Administration[96] (hereinafter NOAA) was introduced with this background but it was not made an independent body, as

---

[90] Lawrence Juda, supra note 10, pp. 164–165.

[91] Section 5 (b) of the Maine Resources and Engineering Act 1966, P. L. 89–454, 17th June 1966.

[92] Commission on Marine Science, Engineering and Resources, *Our Nation and the Sea*, (Washington, D.C.: Government Printing Office, 1969).

[93] Ibid, p. 227–230.

[94] Ibid.

[95] Ibid, p. 227–249.

[96] NOAA was established by Presidential Reorganization Plan No. 4. See 35 Federal Register 15627-15630, 1970.

had been recommended by the Stratton Commission. It was, rather, placed in the Department of Commerce. The creation of NOAA represented a significant step toward a centralised national ocean policy. It did not, however, eliminate the reality that a number of public authorities still maintained important jurisdictional responsibilities regarding ocean governance matters.[97] After this, there were several bills introduced in Congress calling for the establishment of a Department of the Ocean, a Department of the Environment and Oceans or an independent ocean agency.[98]

Unfortunately, due to the energy crisis caused by the Arab oil embargo in the 1970s, the United States Government became more concern with energy requirement and natural resources, rather than with the quality of the environment of the oceans.[99] The fact that the public authorities mobilised to protect existing agencies' jurisdiction was another reason why there was a failure to achieve more substantial organisational change.[100] The consequence of the aforesaid situation would lead to the members of the standing committee of the Congress hesitating and being wary of change, which may have weakened the importance and influence of the committee. Reorganisation also raised anxieties among nongovernmental stakeholders.[101] All of these would, accordingly, suggest that institutional reorganisation need to be carefully and skeptically scrutinised.

A further indicator of the changing ocean and coastal governance approach is evidenced by the establishment of the Congress of the National Estuary Program, in 1987.[102] The programme emphasises the importance of the estuaries ecology, the multiple causes of environmental degradation and monitoring all activities in that area. This programme also applies an ecosystem-based approach to environmental use, protection and sustainability. Similar change with respect to fisheries governance is evidenced by the Magnuson-Stevens Fishery Conservation and Management Act in 1990[103] and 1996.[104] Attention has been paid on the limits of a nature system, the need to protect essential fish habitats from the effects of

---

[97] The General Accounting Office, *Federal Agencies Administering Programs Related to Marine Science Activities and Oceanic Affairs*, GGD-75-61, 25th February 1975, p. 3.

[98] For instance, S. 3889, The Department of the Environment and Oceans Act, 94th Congress, 2nd Session, 30th September 1976 that would have combined, among other public authorities such as NOAA, the Environmental Protection Agency and the Coast Guard; S. 2224, The National Oceanic and Atmospheric Administration Organic Act, 95th Congress, 1st Session, 20th October 1977.

[99] Lawrence Juda, supra note 10, p. 166.

[100] Ibid.

[101] Ibid.

[102] The National Estuary Program was established in accordance with Section 320 of the Clean Water Act Amendments of 1987, Public Law 100-4 and was reauthorised through fiscal year 2005 by the Estuaries and Clean Water Act of 2000, Public Law 106-457.

[103] The Magnuson-Stevens Fishery Conservation and Management Act, as amended through 11th October 1996, Public Law, 101-627.

[104] Ibid above note, Public Law, 104-297.

fishing and non-fishing activities and adopting a wider and systems based approach to fisheries governance.[105]

As evidence of a growing interest in adopting the ecosystem-based approach in ocean governance during the 1990s, the United States, had recognised the need to reconsider the need for institutional and policy adaptation for effective and efficient ocean governance. In addition to the introduction of the Oceans Act 2000,[106] the United States Commission on Ocean Policy was established, in 2001.[107] The Commission aims to review and make recommendations with respect to:

"(A) An assessment of existing and planned facilities associated with ocean and coastal activities including human resources, vessels, computers, satellites, and other appropriate platforms and technologies.

(B) A review of existing and planned ocean and coastal activities of Federal entities, recommendations for changes in such activities necessary to improve efficiency and effectiveness and to reduce duplication of Federal efforts.

(C) A review of the cumulative effect of Federal laws and regulations on United States ocean and coastal activities and resources and an examination of those laws and regulations for inconsistencies and contradictions that might adversely affect those ocean and coastal activities and resources, and recommendations for resolving such inconsistencies to the extent practicable. Such review shall also consider conflicts with State ocean and coastal management regimes.

(D) A review of the known and anticipated supply of, and demand for, ocean and coastal resources of the United States.

(E) A review of and recommendations concerning the relationship between Federal, State, and local governments and the private sector in planning and carrying out ocean and coastal activities.

(F) A review of opportunities for the development of or investment in new products, technologies, or markets related to ocean and coastal activities.

(G) A review of previous and ongoing State and Federal efforts to enhance the effectiveness and integration of ocean and coastal activities.

(H) Recommendations for any modifications to United States laws, regulations, and the administrative structure of Executive agencies, necessary to improve the understanding, management, conservation, and use of, and access to, ocean and coastal resources.

(I) A review of the effectiveness and adequacy of existing Federal interagency ocean policy coordination mechanisms, and recommendations for changing or improving the effectiveness of such mechanisms necessary to respond to or implement the recommendations of the Commission."[108]

While the Commission is considering new proposals as regards the need for interagency coordination, the possible reorganisation of the responsibilities of the public authorities and the balance between marine environment protection and commercial development, the issue of an integrated ocean governance programme

---

[105] Ecosystem Principles Advisory Panel, a report to Congress, *Ecosystem-Based Fishery Management*. This report was mandated by the Sustainable Fisheries Act Amendments to the Magnuson-Stevens Act and available online at www.nmfs.noaa.gov/sfa/EPAPrpt.pdf.

[106] The Oceans Act of 2000, Public Law 106-256, 7th August 2000.

[107] See the United States Commission on Ocean Policy website: http://www.oceancommission.gov/commission/welcome.html last visited date: 07/05/2009.

[108] Section 3 (f) (2) of the Ocean Act of 2000.

remains problematic. As the Commission noted, as pressures and problems in coastal and ocean zones continue to increase, the lack of organisational coordination and coherence in government efforts, as stated by the Stratton Commission, are still very much in evidence.[109] There was a view that a comprehensive national ocean governance programme and the resulting institutional changes might only be achieved after defined goals and objectives were established.[110] The aforesaid statement would indicate a chicken-and-egg scenario, which comes first? To change institutional arrangements or to develop an integrated ocean governance programme? Who is going to develop a comprehensive programme? Or alternately, is it possible for an integrated ocean governance programme to emerge with the present widespread public authorities and the various jurisdictions relating to different issues.

The most recent development is that President Barark Obama has signed an executive order to publish the first National Ocean Policy as well as the establishment of National Ocean Council (NOC). The function of NOC is said to coordinate all the marine related public authorities in order to implement the National Ocean Policy. Inputs not only from the national level but also incorporates the tribal and local levels. The aforesaid effect is considered to be a milestone for more comprehensive and integrated ocean governance approaches.

3.4.2 Canada

Together with the obligations illustrated by UNCLOS, attention has been given to the need for marine environmental protection, marine resources conservation and settling disputes of use. This has led Canada to reassess its ocean governance approaches. Interest in and support for a more innovative, comprehensive and proactive approach to ocean governance increased during the 1990s.[111]

A study launched by the Department of Fisheries and Oceans (hereinafter DFO) stated that there are "pressures on the marine environment highlight the numerous and sometimes competing demands placed on the marine ecosystem. Many of the resulting environmental impacts are the result of unplanned and/or locally driven decisions that have been made without consideration of their wider environmental impacts. This illustrates a need to focus on integrated approaches to marine activities. For example, the multiple or conflicting use of coastal areas, including fishing, aquaculture, tourism, recreation, construction of human habitats, waste discharge, marine mining, and shipping, make the application and success of isolated approaches questionable. For users of the oceans resources to coexist and

---

[109] The United States Commission on Ocean Policy, *Development a National Ocean Policy* (Washington, D. C., September 2002), p. 4.

[110] General Accounting Office, *Need for a National Ocean Program*, GGD-75-97, 10th October 1975.

[111] National Advisory Board on Science and Technology, Committee on Oceans and Coasts, *Opportunities from Our Oceans*, (Ottawa May 1994).

ensure the sustainability of the marine environment, integrated approaches to oceans activities management are critical."[112] Increasingly, it is clear that there is a need for planning and coordination to address the cumulative effects of the operation of the various public authorities. To advance a more integrated ocean governance approach, it is felt that there is an imperative for clear mandates and incentives for cooperation among public authorities and with the private sectors.[113]

From an institutional aspect, the current governmental framework as regards jurisdiction over ocean governance is spread among a number of public authorities. However, the Minister of the Department of Fisheries and Oceans has been granted the overall authority in both the development and implementation of an integrated national ocean policy.[114] Such a prominent status was granted by the Government Organization Act of 1979 which moved the power regarding fisheries management away from the Department of the Environment the Department of Fisheries and Oceans.[115] The Government Organization Act also provides that DFO has residual power over all matters relating to the coordination of ocean governance policies and programmes where there is no appropriate existing legislation.[116] This residual power was confirmed by the Oceans Act[117] and serves to emphasis the role of DFO in the development of a national ocean policy. If other public authorities have been assigned ocean governance responsibilities under a statute, the power to govern the aforesaid activities remain vested in hands of those assigned public authorities, not in DFO.

The position of DFO was further strengthened by incorporating the Canadian Coast Guard within its remit, in 1995. The role of DFO Minister is rather as a coordinator or facilitator, than as a supervisor of other relevant marine public authorities. As regards governmental policies beyond DFO's jurisdiction, the role of DFO Minister is one of coordination and consultation but not overseeing or direction.[118]

It is, however, felt that if Canada's approach is considered to be a failure, it is because there has been a failure to reduce the number of public authorities involved in the marine policy area.[119] As the Parliament's Standing Committee on

---

[112] Canada, Department of Fisheries and Oceans, *Ensuring the Health of the Oceans and Other Seas,* Ottawa 1997, p. 2; see also http://www.sdinfo.gc.ca/reports/en/monograph3/splash.cfm website no longer exist, copy in file.

[113] J. G. Michael Parkes and E. W. Manning, *An Historical Perspective of Coastal Zone Management in Canada* (Ottawa: Department of Fisheries and Oceans, 1998); see also Lawrence Juda, supra note 10, p. 170.

[114] Section 29 and 32 of the Oceans Act, 1996 c. 31, assented to 18th December 1996.

[115] The Government Organization Act of 1979, statutes of Canada 1978–1979, Chapter 13.

[116] Canada, Department of Fisheries and Oceans, *Role of the Federal Government in the Oceans Sector,* (Ottawa, 1997), p. 4.

[117] Section 40 of the Oceans Act 1996.

[118] Section 32, 33 and 34 of the Oceans Act 1996.

[119] International Ocean Institute, *Final Report of the Canadian Ocean Assessment—A Review of Canadian Ocean Policy and Practice* (Halifax: Dalhousie University, October 1996), pp. 82–84.

Fisheries and Oceans concluded, the governance of Canada's oceans was still "becoming increasingly fragmented between different ministers."[120] Accordingly, the Committee suggested that the government confirm that DFO Minister should have primary responsibility for all matters concerned with the governance of Canada's oceans and urged that the Minister exercise the role proactively.[121] Criticism was focused on the fact that DFO's ocean governance activities were based on the pragmatic approach of "learning by doing".[122] The Committee also recommended DFO to prepare an annual report in order to update the implementation of the Oceans Act 1996[123] and perhaps place a certain degree of pressure on DFO to perform.

3.4.3 Australia

The Australia's Ocean Policy noted that, there was a trend in international treaties and instruments toward 'holistic' approaches to ocean governance.[124] The Australian ocean governance approach was, however, sectoral based and characterised by fragmentation of responsibilities within and among different levels of government.[125] This situation is said to be "a tyranny of small decisions and a jurisdictional nightmare, giving rise to multiple, overlaid, uncoordinated and collectively excessive use of resources."[126] Over time, there has been a shift in Australian's thinking regarding ocean governance. Attention is increasingly paid to ecosystem dependent instead of sectoral based approaches. This ecosystem-based approach takes account of the multiple uses of ocean zones and their cumulative effects on ecosystems.[127] The issue which needs to be addressed is that ocean governance arrangements have traditionally been sectoral based and have been developed in the absence of an over all national ocean policy.[128]

---

[120] Canada, House of Commons, Standing Committee on Fisheries and Oceans, *Report on the Oceans Act* October 2001, recommendation 11, available from: www.parl.gc.ca/InfoComDoc/37/1FOPO/Studies/Reports/fopo01-e.htm. last visited date:07/05/2009.

[121] Ibid, recommendation 12.

[122] Ibid, Part II E (2) (A).

[123] Ibid, Part I, recommendation 2.

[124] Commonwealth of Australia, National Oceans Office, *Australia's Ocean Policy—International Agreements*, Background Paper 2, October 1997, available from: www.oceans.gov.au/back-ground_paper_2/title.jsp, website no longer exist, copy in file.

[125] Lawrence Juda, supra note 10, p. 174.

[126] Commonwealth of Australia, National Oceans Office, supra note 124, part 2.1, p1.

[127] Commonwealth of Australia, *Australia's Ocean Policy*, 1998, ISBN 0 642 54580 4, pp. 19–21.

[128] Commonwealth of Australia, National Oceans Office, *Australia's Ocean Policy—Oceans Planning & Management*, Issue Paper 3: Best Practice Mechanisms for Marine Use Planning, September 1997, available from: www.oceans.gov.au/issues_paper_3/title.jsp, website no longer exist, copy in file.

In 1998, after releasing two consultation papers, four background papers and seven issues' papers, the Australian Commonwealth Government published Australia's Ocean Policy.[129] Central to this policy is the development of regional marine plans based on large marine ecosystems.[130] These plans are to be applied to large marine zones, which are determined on a bio-geographic and bio-regional dimension, rather than considering jurisdictional boundaries.[131] Apart from this, Australia's Ocean Policy called for the establishment of a number of institutional arrangements. They are, the National Oceans Ministerial Board of Commonwealth Ministers, the National Ocean Advisory Group, the Regional Marine Plan Steering Committees and the National Oceans Office.[132]

Although Australia's Ocean Policy envisages some institutional change, it does not consider creating a single marine agency. Traditional sectoral based approaches are to be modified rather than replaced. It is also believed that based on the ecosystem approach, integrated ocean governance can be achieved through existing mechanisms and does not require any significant institutional change.[133]

The current practice is that of an Oceans Ministerial Board, chaired by the Environment Minister.[134] The Marine Division of the Department of the Environment and Heritage provides central coordination and policy advice to the Australian Government on the marine environment, including the implementation and further development of Australia's Oceans Policy.[135] The division is further divided into three branches, namely, the National Oceans Office, the Marine Conservation Branch, the Marine Environment Branch.[136]

The National Oceans Office is the lead Australian Government agency responsible for the implementation of Australia's Oceans Policy; the Marine Conservation Branch has a range of responsibilities for marine environment policy development and implementation. It has primary responsibility for the development and implementation of integrated oceans management and for Marine Protected Areas. The branch also coordinates marine policy throughout the Department and provides high level policy advice on a range of marine issues; the Marine Environment Branch undertakes activities at regional, national and

---

[129] Joanna Vince, supra note 80, p. 422.

[130] Ibid, p. 420.

[131] Commonwealth of Australia, The Marine Science and Technology Plan Working Group, *Australia's Marine Science and Technology Plan*, June 1999, p. 25 and 40, available from: www.isr.gov.au/science/marine/marineplan, website no longer exist, copy in file.

[132] Lawrence Juda, supra note 10, p. 175–176.

[133] Commonwealth of Australia, National Oceans Office, *Report of the Ministerial Advisory Group on Oceans Policy*, at section 3.1, "Institutional Arrangements" available from: www.oceans.gov.au/magaop_report/title.jsp, website no longer exist, copy in file.

[134] Joanna Vince, supra note 80, p. 424; Malcolm MarGarvin, supra note 28, p. 8; Lawrence Juda, supra note 10, p. 175.

[135] See Australian Government, Department of the Environment and Heritage http://www.deh.gov.au/md/index.html, website no longer exist, copy in file.

[136] Ibid.

international levels on a range of issues, including responsibility for protected and migratory marine species and managing Australia's obligations under UNCLOS.[137]

It is suggested that the differences within current government level, federal and state, tend to be underestimated in terms of the effects of lacking of coordination between federal and state governments.[138] What is clear is that attempts to develop an integrated ocean governance will face difficulties, if federal and state governments and the general public cannot achieve a consensus regarding ocean governance issues.

3.4.4   Summary

There are two possibilities for changing institutional structure in the context of ocean governance. Firstly, to provide sufficient, evaluated marine policy information to the highest levels of the executive and legislative branches of government. The aim is to ensure that a single public authority is continuously assessing relevant ocean governance activities, both at national and international levels. This implies definite assigned levels of responsibility and authority to look after overall national ocean policy.[139]

Secondly, to unify and harmonise the various mission-related activities of the public authorities which have relevant ocean governance programmes.[140] The intention is to bring together and concentrate the efforts of the various public authorities working in this area. The purpose is to avoid duplication and waste through more wisely governing the available resources.[141]

There are two ways of achieving the aforesaid goals. At one end of the scale might be a new ocean super-agency.[142] The proposal for a 'World Ocean Authority' can be seen as an extreme version of this strategy.[143] At the other end would be the continuation of the status quo with nothing more than a slight improvement in the existing consultative arrangements for coordination of national ocean policy. The extremes of this might be "from doing too much to doing too little".[144] The primary consideration for changing infrastructure is to give decision

---

[137] Ibid.

[138] Marcus Haward, supra note 79, p. 531–532.

[139] Don Walsh, "Some Thoughts on National Ocean Policy: The Critical Issue", *San Diego Law Review* 13 (1976):619.

[140] Ibid.

[141] Ibid.

[142] Arild Underdal, supra note 15, p. 167.

[143] The Cousteau Society, *Basic Principles for a Global Ocean Policy*, Principle 10, January 1979.

[144] Don Walsh, supra note 139, p. 619–620.

makers the tools to solve the ocean governance problems. Infrastructure need not take the form of an organisation but may be the appropriate structure to make and implement rule and utilise the appropriate resources efficiently.[145]

3.5 Conclusion

There is an increasing recognition of the problems posed by sectoral approaches to ocean governance. An integrated marine policy seems to be a solution to the aforesaid problem. The experiences of the United States, Canada and Australia suggest that decisions taken in different areas, without sufficient reference to actions by others, seem to create significant problems. There is also a clear evidence of an interest in resolving and rethinking such difficulties and the way in which ocean governance has been approached. There is a desire, in all three States, to develop approaches to the marine environment that are proactive and systematic rather than reactive and specific use-based.

In the United States, change has been made through programmatic and legislative modifications which encourage a broader participation in the decision-making processes. However, there is no overriding national ocean policy but rather a number of marine policies based on particular statutes. There is no single ocean agency but NOAA, has accentuated indirect policy methods rather than direct institutional change through reorganisation.

In Canada, the Oceans Act has legislated for a broad ocean policy and indicated its expectations. Nonetheless, it is lacking an institutional mechanism for change and implementation of the said policy. As Lawrence Juda observes "the Oceans Act promises more than it delivers."[146]

In Australia, a new institutional structure has been established based on Cabinet decisions, rather than parliamentary acts. This concept, to date, receives the support of the Australian state governments whose cooperation is essential to achieve good ocean governance.

It is not the intention of this chapter to provide a specific solution to the problems of ocean governance. Attempt is to stimulate thought and perhaps initiate dialogue which can help the decision makers to acquire a broader and more comprehensive view of their responsibilities. Bearing the review of the three State practices, it is not surprising that change has not been more rapid although change is occurring. In light of this, alterations in decision makers' attitudes, law and institutional arrangements can be expected in the future.

---

[145] Robert L. Friedheim, "Ocean Governance at the Millennium: Where We Have Been—Where We Should Go", *Ocean & Coastal Management* 42 (1999):750.

[146] Lawrence Juda, supra note 10, p. 179.

Chapter 4
The Governance of Marine Resources

Abstract This chapter discusses the trade-off between the economic development and environmental protection in terms of governance of marine resources. Attention, then, being paid to sustainable development of marine resources and distinct marine natural resources into renewable resources and exhaustible resources. The paper also discusses how economics activities have impacted on marine resources and in what way these issues might encapsulate the legal concern. This chapter concludes that a holistic governance approach which would be greatly facilitated by the development of a comprehensive and coordinated national oceans policy, similar to those already adopted by a number of States. Such policies would provide guiding principles and detailed programmes to enable and encourage all government departments dealing with oceans issues, to consult each other and to coordinate their work.

Keywords Sustainable development · Marine resources · Renewable resources · Exhaustible resources · Sustainable economics

4.1 Introduction

The radical precondition for effecting human activities not only one, ethics, economics, science and law, all play important roles. When issue of 'sustainable development' is discussed, all of the aforementioned elements should be equally considered. If the question of 'which particular discipline has contributed to the problems of unsustainable development' being asked. 'Economics', believe will lead the debate. Therefore, when issue of governance of marine resources is discussed, the trade-off between the economic development and environmental protection should be considered.

Y.-C. Chang, *Ocean Governance*, SpringerBriefs in Geography,
DOI: 10.1007/978-94-007-2762-5_4, © The Author(s) 2012

Development relies on resources. Resources can be further divided as artificial resources (e.g. factory equipments, cars, cargo ships and human resources) and natural resources (e.g. fishes, forest, fossil fuels).[1] Artificial resources are wholly dominated by human activities and can always be replaced. Therefore, this chapter does not intend to discuss the governance of artificial resources, rather it will concentrate on the natural resources. Natural resources are often categorised as renewable and exhaustible.[2] The contemporary development of marine policy has moved from marine resources exploitation to marine resources conservation and protection.[3] Renewable and exhaustible resources are central to the issues of the sustainable development of marine resources and it is necessary that these be considered. Their centrality to the problem of sustainability makes it appropriate to consider them as part of the section on the sustainable development of marine resources.[4]

This chapter below will start with debating on economics development and sustainability. Attention, then, being paid to sustainable development of marine resources (artificial resources are outwith the scope of this discussion) and distinct marine natural resources into renewable resources and exhaustible resources.[5] The paper will finally describe and analysis how economics activities have impacted on them and in what way these issues might encapsulate the legal concern.

4.2 Economic Development and Sustainability

Economic activities affect the environment in many diverse ways. In producing and consuming goods and services societies modify the chemical compositions of the atmosphere, soils, fresh waters and ocean; alter the vegetation cover of the land and the diversity of wildlife inhabiting both land and water. Some of these environmental modifications are intentional, such as those achieved through processes of agriculture and forestry, from urbanisation and the construction of social

---

[1] As Blueprint 3 mentioned "the bequest to the next generation comprises a 'mix'of man-made and 'natural'capital. It is the aggregate quantity that matters and there is considerable scope for substituting man-made wealth for natural environmental assets." For detail please see David Pearce, *Blueprint 3: Measuring Sustainable Development* (Earthscan Publications Limited, 1993), p. 16.

[2] R. Kerry Turner, David Pearce and Ian Bateman, *Environmental Economics* (The Johns Hopkins University Press, 1993), p. 205.

[3] Sam Bateman and Dick Sherwood, *Ocean Management Policy: The Strategic Dimension* (University of Wollongong, Australia, 1994), p. 11.

[4] John Bowers, *Sustainability and Environmental Economics: An Alternative Text* (Addison Wesley Longman Limited, 1997), p. 175.

[5] Renewable and exhaustible resources are central to the issues of sustainable development and necessary to be considered. Their centrality to the problem of sustainability makes it appropriate to consider them as part of the section on sustainable development. For detail please see John Bowers, ibid, p. 175.

infrastructure like roads, factories and power stations. Other environmental impacts are incidental and often the unintentional by products of economic activities. These include discharge of wastes from industry and domestic living, impacts of tourism and spill over effects of urbanisation.[6]

If broadly look at international economic patterns, the environmental crises are linked to economic development prevalent in the North and in the South and in the economic relationships between North and South.[7] The global environmental problems generated mainly in the North (e.g. greenhouse gases), are seen by many as threatening the ability of the South to develop or not allowing the South 'sufficient and adequate' environmental space for its future development.[8] The international trade system has emphasised the export of natural resources from the South to the North and the import by the South of manufactured goods from the North, contributing to overexploitation of the South's natural resource heritage.[9] This circumstance reflects to two international trade patterns. First, the value of trade in manufactured goods grew at a faster rate than in primary products other than fuel and a growing number of developing countries have emerged as major exporters of such goods. Manufactured goods now account for twice the value of developing countries' non-oil exports. Secondly, the industrialised market economies have come to depend more on fuel imports from developing countries.[10] Those patterns also show that the non-living resources such as fuels and minerals, as well as manufactured goods, are now far more important than tropical products, fish and fish products and other agricultural materials in the flow of primary products from developing to developed countries.

Generally speaking, economics has three important roles in debates on sustainable development:

1. To examine the costs and benefits of achieving particular sustainability objectives, such as limiting the emissions of greenhouse gases or conserving biodiversity;
2. To assess the effectiveness of alternative policy instruments for meeting those objectives. An understanding of economics is required in order to predict how people will behave when faced with different types of taxes and subsidies or legally enforced environmental standards;
3. To assess the costs and benefits of alternative policy instruments.[11]

---

[6] John Bowers, supra note 4, p. 3.

[7] Biliana Cicin-Sain and Robert W. Knecht, *Integrated Coastal and Ocean Management: Concepts and Practices* (Island Press, 1998), pp. 81–82.

[8] Ibid.

[9] Ibid, pp. 82–83.

[10] World Commission on Environment and Development, *Our Common Future* (Oxford University Press, 1987), pp. 78–79.

[11] John Bowers, supra note 4, p. 4.

The aim of those three functions is to put a monetary value on the environ-mental effects of economics decisions and to provide a framework for comparing the environmental losses with economics gains. This is essentially the theory of the operation of markets.[12] It explains how markets distribute resources and states the conditions under which freely functioning markets will maximise society's wel-fare.[13] Thus, if economics activities are not sustainable, then this is because markets are failing to make adequate provision for the future.

4.3 Sustainable Development of Marine Resources

The ocean, seas, islands and coastal areas form an integrated and essential compo-nent of the earth's ecosystem and are critical for global food security, for sustaining economic prosperity and the well-being of many national economies, particularly in the developing countries.[14] Recent estimates by the Food and Agricultural Organi-zation (hereinafter FAO) indicate that at least 60% of world fisheries are either fully exploited or over-fished.[15] While the proportion of stocks that are over-fished is now growing at a slower rate than in the mid-1990s, the FAO estimates that the total marine catches from most of the main fishing areas in the Atlantic Ocean and some in the Pacific Ocean, may have reached their maximum potential years ago and sub-stantial total catch increase from those areas are unlikely.[16]

The implications of such developments for global food security and income generation for both present and future generations were assessed by the Commission on Sustainable Development, when it reviewed the oceans and seas at its seventh session.[17] The Commission recommended that particular priority should be given to the conservation, integrated and sustainable management and sustainable use of marine living resources, including the ecosystems of which they are a part.[18]

Significant progress has been achieved over the past decade in promoting an integrated approach to coastal and ocean governance.[19] At the beginning of the 1990s, efforts were limited to a relatively few countries and were concentrated at

---

[12] Ibid.

[13] Ibid, p. 5.

[14] United Nations Report, Plan of Implementation of the World Summit on Sustainable Development, Johannesburg, 26 August–4 September 2002, p. 16.

[15] FAO Fisheries Department, *Fishery Statistics: Reliability and Policy Implications*, available from: http://www.fao.org/DOCREP/FIELD/006/Y3354M/Y3354M00.HTM last visited date: 13/01/2008.

[16] Ibid.

[17] United Nations Report of the Secretary-General, Economic and Social Council, Ocean and Seas, 30 April–2 May 2001, E/CN.17/2001/PC/16, p. 3.

[18] United Nations Report of the Secretary-General, Economic and Social Council, Ocean and Seas, 30 April–2 May 2001, E/CN.17/2001/PC/16, p. 3.

[19] Biliana Cicin-Sain and Robert W. Knecht, supra note 7, pp. 39–64.

the local and community levels, especially in isolated regions. Today, however, an increasing number of States are operating some level of coastal and ocean planning, including small island developing States.[20] Institutional development and legal codification of these new practices at the level of the State are also increasing.[21] The Rio Declaration states that since human behaviour is the major contributor to the degradation of the marine environment, then clearly, it is the responsibility of human beings to take such action as is required to improve and sustain the said environment.[22]

When there is a discussion of the term sustainable development of marine resources, the concept may be defined as the "development that meets the needs of the present without compromising the ability of future generations to meet their own needs."[23] Despite the above definition, there are some in the debate regarding sustainable development who feels that sustainable development can be achieved if there is a balance between economic goals and protection of marine environment. The Brundtland Report contains two key concepts: "the concept of 'need', in particular the essential needs of the world's poor, to which overriding priority should be given; and the idea of limitations imposed by the state of technology and social organization on the environment's ability to meet present and future needs."[24] In the report, the sustainable development of marine resources was largely defined as a trade-off between economic development and the protection of marine environment. This appears to be a competition between the two factors, because many of the earth's resources are finite and in relation to these, it is impossible to achieve growth, social justice and environmental benefits simultaneously.[25] Some environmentalists argue that sustainable development in its narrowest definition is 'economy v. environment' and any balance is impossible because economic interests will always predominate. Because environmental protection standards cannot be easily seen as a threat to marine environment and this coupled with strong economic growth. In light of the current environmental challenges, this chapter uses the term 'Sustainable Economic'[26] to mean economic

---

[20] United Nations Report of the Secretary-General, Economic and Social Council, Ocean and Seas, 30 April–2 May 2001, E/CN.17/2001/PC/16, p. 2.

[21] Ibid.

[22] Principal 1 of the Rio Declaration reflects a shift towards an anthropocentric approach to environmental and development issues, declaring that hu***man beings are "at the centre of concerns for sustainable development", and that they are "entitled to a healthy and productive life in harmony with nature"; this falls short of recognising a right to a clean and healthy environment. For detail please see Philippe Sands, *Principles of International Environmental Law I* (Manchester University Press, 1995), pp. 49–50.

[23] Report of the World Commission on Environment and Development, *Our Common Future*, p. 8.

[24] Ibid, p. 43.

[25] J. Alder and D. Wilkinson, *Environmental Law and Ethic*, (London Macmillan, 1999), p. 141.

[26] Joseph R. Des Jardins, *Environmental Ethics—An Introduction to Environmental Philosophy*, Third Edition (Wadsworth Group, 2001), p. 60.

development which takes into consideration the effects of such development on the marine environment. The main research question, then, can be presented as 'Does the current ocean governance mechanism practice sustainable economics?'

The point of this brief discussion is to emphasise the inescapable fact of the interdependence between economic development and governance of marine resources. The interdependence of almost all States is now so great and their patterns of production and commerce are so interwoven, that it makes little sense to try to solve the problems of sustainable development on a national or regional level; it can only be achieved, if at all, on a global basis. As the Brundtland Commission pointed out, *"Until recently, the planet was a large world in which human activities and their effects were neatly compartmentalized within nations, within sectors (energy, agriculture, trade), and within broad areas of concern (environmental, economic, social). These compartments have begun to dissolve. This applies in particular to the various global 'crises' that have seized public concern, particularly over the past decade. These are not separate crises: an environmental crisis, a development crisis, an energy crisis. They are all one."*[27]

The overriding concern is the fact that international governmental organisations, such as the United Nations, not to mention international non-governmental organisations (NGOs), remain politically quite weak, when compared to national regimes. The former lack both the resources and regulatory authority to take decisive action in transnational issues. To great extent, they are often relegated to an advisory role, with little more than the force of moral suasion,[28] which weakens their ability to deal with the marine issues as a whole. The objects of governance of marine resources will not be achieved unless all planet members have the same objectives and a detailed knowledge of their surrounding oceans. They would also have to obey the international legal instruments, otherwise the sustainable development of marine resources will remind an unachieved mission.

4.4 Marine Natural Resources

From an international legal aspect, different ocean exploration or exploitation activities are affected by different international instruments such as the 1972 Convention on the Preservation of Marine Pollution by Dumping of Waste and Other Matter,[29] the 1992 Convention on Biological Diversity,[30] the 1978

---

[27] World Commission on Environment and Development, *Our Common Future*, p. 4.

[28] Edited by Frank Fischer and Michael Black, *Greening Environmental Policy: The Politics of a Sustainable Future* (Paul Chapman Publishing Ltd, 1995), p. xii.

[29] Convention on the Preservation of Marine Pollution by Dumping of Waste and Other Matter (London, Mexico City, Moscow, Washington DC) 29 December 1972, came into force 30 August 1975; 1046 UNTS 120.

[30] Convention on Biological Diversity (Rio de Janeiro) 5 June 1992, came into force December 1993; 31 ILM 822 (1992).

Convention on Future Multilateral Co-operation in the Northwest Atlantic Fisheries[31] and the 1989 Kuwait Protocol Concerning Marine Pollution Resulting from Exploration and Exploitation of the Continental Shelf.[32] Since this chapter is concerned with the governance mechanism, inevitably it will touch upon issues in relation to marine renewable and exhaustible resources. The discussion below considers the renewable, exhaustible resources and asks how they are governed and how they should be governed.

4.4.1 Renewable Resources

Renewable marine resources are natural plants and animals that are regenerated themselves and can be exploited by humans. The notion excludes plants and animals that are domesticated and whose populations are subject therefore to direct human control. Thus, the renewable marine resource theory does not apply to fish farming or aquaculture. It applies to fishing, rather than fish farming and to any animal that is hunted. It would also apply to the collection of wild plants.[33]

In contrast to property rights, as regards renewable resources, the concept of common property applies, which means that anyone, or any of a large number of people, is able to enter the industry and exploit the stock. It is often believed that the notion of common property of renewable resources is the major cause of over-exploitation, the best example being fishing.[34] The risks associated with the open access solution are often summarised as the *"tragedy of the commons."*[35] Open competition will naturally lead to a situation as though no property rights exist. Even if such rights do exist, their claim to ownership can easily be challenged or ignored and therefore, with regard to this 'State ownership' might be one of the solutions. For State ownership to operate efficiently, the State must be able to monitor the use of resources, establish acceptable rules of use by individuals, community and enforce those rules.[36] The open-access problem not only exists for fishing in international waters in the absence of international agreements on fishing limits, it also exists in territorial waters, unless there are restrictions on who can fish and how much fish they may take.

---

[31] Convention on Future Multilateral Co-operation in the Northwest Atlantic Fisheries (Ottawa) 24 October 1978, came into force 1 January 1979; Cmnd. 7569.

[32] Kuwait Protocol Concerning Marine Pollution Resulting from Exploration and Exploitation of the Continental Shelf (Kuwait), 29 March 1989, came into force 17 February 1990. Available from: http://feaapp.fea.gov.ae last visited date:7/26/2004.

[33] John Bowers, supra note 4, p. 183.

[34] Edward B. Barbier, *Economics, Natural-Resource Scarcity and Development—Conventional and Alternative Views* (Earthscan Publications Limited London, 1989), pp. 66–67.

[35] R. Kerry Turner, David Pearce and Ian Bateman, supra note 2, p. 210. John Bowers, supra note 4, p. 189.

[36] John Bowers, supra note 4, p. 218.

With regard to open access, if economic factors lead to over-exploitation, the only factor that can prevent the species from being driven to extinction is the rise in costs resulting from declining stocks, which may make fishing temporarily uneconomic and allow the species capacity to recover. Even so, rises in stocks will cause those who have left the market to re-enter it again, so that the fish population fluctuates considerably and in addition, the species is always at risk from technical progress in fishing methods making fishing profitable, even at low population levels.[37] With many renewable marine resources, current population levels are extremely unstable levels and the population faces the risk of extinction, which is incompatible with sustainability. This is the case with many pelagic fish: haddock, cod and notably the North Sea herring.[38]

The traditional legal solution for this problem is for States to set up property rights in fisheries and to control entry to them or to control catch levels by imposing quotas, supported by the domestic laws. For example, in the United Kingdom, new legislation has progressively introduced greater "byelaw-making powers for use in a nature conservation context, increasing the scope for marine site protection."[39] The overall effectiveness of site management frameworks is, however, questionable.[40] The only route that seems to be currently available is to protect fish stocks, would be prohibited fishing on certain areas and restricting catches in other areas.[41] However, at the moment, such restriction can only be agreed on a case-by-case basis through the Council of Fisheries Ministers.[42]

Reflecting on the call for overall governance, even if the public authority realises the serious of circumstances, the public policy still needs to equitably and inclusively form a compromise among all of the different interests in society. The participation of stakeholders within the decision-making processes will seal the gaps between the ideal and the reality. The other strategy to increase the yields of renewable marine resources, without endangering stock levels, is fish farming.

Farmed species are not renewable marine resources, since the reproduction of the species is then within human control. The farming of salmon, trout, carp and other species allows high levels of consumption, without endangering the wild species. Finally, investing in research on techniques of cultivation of threatened renewable marine resources can form part of a sustainability strategy for renewable marine resources.

---

[37] John Bowers, supra note 4, p. 189.

[38] The Independent Newspaper, Wednesday 8 December 2004, p. 1,4, 30; John Bowers, supra note 4, p. 189; Royal Commission on Environmental Pollution, Twenty Fifth Report, *Turning the Tide—Addressing the Impact of Fisheries on the Marine Environment*, 7 December 2004, par. 3.14–16.

[39] Royal Commission on Environmental Pollution, Twenty Fifth Report, *Turning the Tide—Addressing the Impact of Fisheries on the Marine Environment*, par. 4.79.

[40] Ibid.

[41] Ibid, par. 4.78.

[42] Ibid.

4.4.2 Exhaustible Resources

Exhaustible resources are primary materials such as minerals and fossil fuels extracted from the earth.[43] They are created from a long term geological process and can be regarded as fixed in quantity.[44] Due to the limitations of technology, the total quantity in existence may not yet be known. Because they are fixed in quantity and cannot be reproduced, the more extract and use today, the less there will be for future generation.[45] Thus, the consumption of a quantity of an exhaustible resource at any time carries an opportunity cost, i.e. the value of consuming that resource in the future. This opportunity cost of consumption is usually given the name *'user cost'*.[46] The nature of exhaustible resources means that the important economic question is when to use them. Leaving aside issues of recycling, the use of coal, oil, metal ores, limestone and so on, reduces the quantity that can be consumed by future generations.

The problem that exhaustible resources pose for sustainability is that they are exhaustible. Although recycling and reuse, depending on technology, is possible to a limited degree, use at one period denies their use in the future. Thus, it is impossible for the current generation to leave the same quantity of exhaustible resources for future generations. To achieve the sustainability of exhaustible resources, there are several approaches which depend on appropriate technical capability:

1. "Improvements in efficiency in primary use;
2. Efficiency in recycling;
3. Efficiency in extraction;
4. Replacement of the exhaustible materials by renewable resources."[47]

The first three options listed above extend the use of exhaustible resources, increasing their expected life. The last alternative is the only way to achieve the requirement of sustainable development, with the use of wind, wave and solar power but this ideal requires further technology development.

Exhaustible resources are not necessarily reserved to maritime zones but the majority of them lie in ocean areas. With regard the governance of marine resources, the approaches mentioned above in relation to the exhaustible resources might become one of the policy options.

---

[43] John Bowers, supra note 4, p. 176.

[44] Robert I. Williams, 'Developments in the Economic Theory of Non-Renewable Resources: Implications for Policy', Harvard Environmental Law Review, vol.2 (1977), p. 562; R. Kerry Turner, David Pearce and Ian Bateman, supra note 2, p. 205.

[45] R. Kerry Turner, David Pearce and Ian Bateman, supra note 2, p. 221.

[46] The notion of user cost does not exist for conventional reproducible resources since the consumption of an amount now does not reduce the quantity that can be consumed in the future.

[47] John Bowers, supra note 4, p. 181.

The allocation of available resources is generally based upon previously established needs, rather than on the potential wider needs of the society. Failure to recognise the wider needs may then lead to the price of exhaustible resources would be too low and the resources would be extracted too rapidly. This once again shows the interdependency of the various decisions which have to make about the future.

4.5 Conclusion

Legally, at international level the 1982 Convention on the Law of the Sea[48] together with other law of the sea instruments are the cornerstones for setting up the standards to examine State practice. At domestic level, two fundamental concerns should be addressed when further examining the current State practice: (1) the degree of concentration or diffusion of power and authority among national-level government institutions; (2) the division of authority between national and sub-national levels of government. Since the problems of ocean space are closely interrelated and need to be considered as a whole. This may require to propose a holistic approach which would be greatly facilitated by the development of a comprehensive and coordinated national oceans policy, similar to those already adopted by a number of States. For example, the establishment of a leading public authority to overseeing the ocean governance issues such as Fisheries and Oceans, Canada; National Oceans Office, Australia; Commission on Ocean Policy, United States.

The above mentioned policies would provide guiding principles (e.g. precautionary principle; the concept of good ocean governance[49]) and detailed programmes (e.g. integrated coastal zone management; ecosystem system based approach) to enable and encourage all government departments dealing with oceans issues, to consult each other and to coordinate their work.[50] The result should be not only more effective management of the oceans at national level but also a uniform and consistent national position at regional and global levels. The consequence of this would foster better cooperation among States, as well as between international organisations addressing oceans issues, potentially leading to more integrated and effective ocean governance at global level.[51]

---

[48] The Untied Nations Convention on the Law of the Sea, Montego Bay, 10 December 1982, came into force 16 November 1994, 21 ILM 1245 (1982).

[49] Yen-Chiang Chang, 'Good Ocean Governance', in *Ocean Yearbook* (Canada: Brill), volume 23.

[50] Report and the Secretary-General, Economic and Social Council, Unite Nations, Ocean and Seas, E/CN.17/2001/PC/16, 14 March 2001, p. 5.

[51] Ibid.

Chapter 5
Ship Recycling: An Overview of the 2009 Hong Kong International Convention on the Safe and Environmentally Sound Recycling of Ships

Abstract This chapter discusses the '2009 Hong Kong International Convention on the Safe and Environmentally Sound Recycling of Ships', from its historical background, structure and enforcement. The 2009 Hong Kong Convention establishes control and enforcement instruments in relation to ships' recycling such as Flag State obligations, Port State control rights, obligations of a Party and its recycling facilities under its jurisdiction, communication and exchange of information procedure, reporting system upon completion of recycling, plus detection of violations and auditing system. The Convention, however, also contains some deficiencies. This chapter concludes that it is these deficiencies that will eventually influence the final acceptance of this Convention by the international society.

Keywords Ships recycling · Marine environmental protection · International maritime organisation

This article was earlier published as Ship Recycling and Marine Pollution, Yen-Chiang Chang et al., Marine Pollution Bulletin 60 (2010): 1390–1396.

5.1 Introduction

The disposal of the ships, when they reach the end of their economic life has great significance for the continual renewal of the merchant marine fleet[1] and for sustainable development.[2] The ship recycling facilities, however, also have further negative effects, in terms of the environment and occupational health and safety. 90–95% of international commercial goods are transported by sea, as a result of cost efficiency. Shipping is an international activity, since ships sail throughout the world and is the most important link in the world manufacturers' global logistical chain. Thus, the shipping industry represents the smallest part of a product's cost, in turn making the trade viable.

Nonetheless, the shipping industry poses potentially negative impacts on the marine environment and can incur some economic disadvantages. In addition, if there is no appropriate integrated system for the recycling or reusing of ship related steel, machines, auxiliaries and even furnishings, such materials will remain unused and useless to the economy, at the end of a ship's life cycle. In this respect, the ship recycling facilities contribute to sustainable development and represent the environmentally friendly way of disposing of ships[3] and economically integrating the life chain of ships. On the other hand, particularly in the European Union (EU) and the United States of America (USA) because of the strict regulations relating to environmental issues and occupational safety and health issues, the ship recycling costs are comparatively higher than the costs in Asian. Thus, ship recycling facilities in the EU and the USA are not economically viable.

At the end of a ship's life cycle, the ship contains various recyclable materials but also a range of hazardous and toxic substances.[4] In Europe and Member States of Organization for Economic Co-operation and Development (OECD), materials which contain hazardous and toxic substances are subject to monitoring and their disposal is strictly regulated. Most of those substances are defined as hazardous and toxic under the existing 1989 Basel Convention on the Control of Transboundary Movements of Hazardous Wastes and their Disposal (1989 Basel Convention).

The reality is that currently, the global shipping industry relies on the developing countries to dispose of decommissioned ships through the process of ship recycling. As a result, the ship recycling industry avoids the burden of complying

---

[1] Ian White and Fionn Molloy, "Ships and the Marine Environment," Speech delivered at the Maritime Cyprus 2001 Conference, Limassol, Cyprus, 23–26 September 2001.

[2] O. Sundelin, *The scrapping of vessels —an examination of the waste movement regime's applicability to vessels destined for scrapping and potential improvements made in the IMO Draft Convention on Ship Recycling* (Master's Degree Easy, University of Gothenburg, Sweden, 2008), p. 10–12.

[3] Nikos Mikelis, "Developments and Issues on Recycling of Ships," Speech delivered at the East Asia Seas Congress, Haikou City, Hainan Province, PR China, 12-16 December 2006.

[4] K. Krause, "End-of-life Ships–Linking European Maritime Safety to Occupational Safety on Asian Scrap Yards," in R. Allsop, J Beckmann and G. M. Mackay (ed.), *ETSC Yearbook 2005, Safety and Sustainability* (Brussels, European Transport Safety Council, 2005), p. 76–80.

with the high cost standards in the developed countries, in order to manage the hazardous wastes involved in decommissioning. Subsequently, the occupational safety and health issues arise, particularly in association with dismantling of beached ships in India, Bangladesh and Pakistan.[5]

The ship recycling workers live under the threat of occupational accidents, due to the inherent risks of ship dismantling. In addition, particularly, in the above mentioned States, most of the workers do not wear protective equipment, such as helmets or masks, neither are there warning signs of danger. Most of the workers have no occupational training in working with blowtorches or with the hazardous substances involved. Furthermore, many do not wear safety goggles to protect their eyes from the sparks. The ship paint and coatings cladding the hull may be flammable and/or may contain toxic ingredients such as polychlorinated biphenyls (PCBs), heavy metals and pesticides such as tributyl tin (TBT).[6] Toxic fumes are released during the blowtorch cutting process and after, whilst the paint and coatings may continue to smoulder. Workers who are using the cutting torches routinely inhale the toxic fumes, for example, from the steel coated with toxic paints.

The concept of environmentalism has been influencing the development policies of the developed countries and the international organisations, in particular, the International Maritime Organization (IMO) —the United Nations dedicated legislator of the global shipping industry—since the catastrophic maritime catastrophes took place in the 1960 s.[7] With regard to this, the IMO has been challenged to establish a globally applicable and comprehensive maritime environmental legal system, in order to achieve the goal of sustainable development. Taking into consideration shipping activities and sustainable development, the maritime environmental legal system comprises three periods, these being;

- The ship construction period,
- The operation and utilisation period of ship for any purpose at sea and
- The ship recycling[8] period.

In the first instance, the IMO aims to develop goal-based construction standards for new ships and determine the fundamental standards of new ship construction. The notion of "goal based ship construction standards" was introduced by the

---

[5] Peter Rousmaniere, "Shipbreaking in the Developing World: Problems and Prospects," *International Journal of Occupational Environmental Health* 13 (2007): 359–368.

[6] Louise de La Fayette, "The Protection of the Marine Environment—1999," *Environmental Policy and Law* 30 (2000): 51–60.

[7] C. K. Hadjistassou, *International Maritime Organization: Rethinking Marine Environmental Policy* (Master's Degree Easy, Massachusetts Institute of Technology, United States, 2004), p. 20–22.

[8] Ship recycling, ship breaking, ship scrapping, ship dismantling and ship demolition are not well-defined terms. With due respect to the terms which are used in the Hong Kong Convention, the authors prefer to use the term of 'ship recycling' in this study. Another reason for using the term 'ship recycling' is related to the comprehensive content of recycling process, rather than breaking, scrapping, dismantling or demolition.

IMO at its 89th session of the Council in November 2002. The recommendations on 'goal based ship construction standards' indicates that the IMO should develop initial ship construction standards which would permit innovation in design but ensure that ships are constructed to a suitable standard and if properly maintained, they should remain safe for their entire economic life. In addition to that, the standards would also have to ensure that all parts of a ship could be easily accessed, in order to permit proper inspections and for ease of maintenance. With regard to these considerations, the IMO Assembly adopted the strategic plan for the 2004–2010 periods and Resolution A. 944(23), which stated that, 'the IMO would establish goal based standards for the design and construction of new ships'.

Subsequently, the IMO, taking into consideration of ship operation period, adopted international legal instruments which aim to challenge vessel sourced marine pollution. The documents which are adopted by the IMO comprise a broad range of marine pollution related issues, including prevention of pollution by oil, carriage of chemicals by ships, prevention of pollution by harmful substances carried by sea in packaged form, prevention of pollution by sewage, prevention of pollution by garbage, prevention of air pollution and prevention of marine pollution by dumping of waste.

Nevertheless, although the IMO aims to establish a global maritime environmental legal system, legal actions relating to the ship recycling period are still in their infancy, when compared to ship construction and ship operation periods. The environmental concerns relating to ship recycling facilities have been taken into account since the 1980s. The first noteworthy attempt to consider ship recycling facilities is the 1989 Basel Convention. Having said that, the 1989 Basel Convention was not entirely appropriate to challenge all risks and problems arising from ship recycling facilities. The 1989 Basel Convention briefly mentions the transportation of hazardous materials, however, it does not indicate detailed rules for the recycling process. With regard to these considerations, the IMO took ship recycling and its environmental impact onto the IMO's agenda. Since 2003, the IMO has adopted the Guidelines and Circulars for the purpose of achieving a green ship recycling industry. In addition, joint efforts with the IMO, the International Labour Organization (ILO) and the Conference of Parties to the 1989 Basel Convention, were also made to establish a Joint Working Group on ship scrapping. Finally, in May 2009, the Hong Kong International Convention for the Safe and Environmentally Sound Recycling of Ships was adopted, which was aimed at ensuring that ships, when being recycled after reaching the end of their operational lives, do not pose any unnecessary risk to human health and safety or to the environment.

5.2 The Historical Background

Taking into consideration the environmental issues, ship recycling facilities and their consequences were first brought to the IMO Marine Environment Protection Committee (MEPC) at its forty second session (MEPC 42), in 1998. Since then, the Committee has agreed that the IMO has a dominant role to play in regulating ship recycling facilities. The IMO's role comprises technical and legal aspects, such as preparation of the proceedings for a ship before recycling commences and a co-ordinating role as regards the ILO and the 1989 Basel Convention, during the recycling period. At the MEPC 47 session, the Committee agreed that the IMO should develop recommendatory guidelines, to be adopted by an Assembly resolution.

The MEPC 49 conducted the 'IMO Guidelines on Ship Recycling', which were adopted at the twenty-third regular session of the Assembly by Resolution A. 962(23), on 5th December 2003. The above mentioned Guidelines comprise identification of potentially hazardous materials, procedures for new ships relating to ship recycling, procedures for existing ships relating to ship recycling, preparations for ship recycling, the role of stakeholders and other bodies and technical co-operation issues.

It was agreed in the MEPC 53, that the IMO should develop a new mandatory system relating to ship recycling activities. The system should aim to establish legally binding and globally applicable regulations, for ship recycling facilities. With regard to these considerations the twenty-fourth regular session of the IMO Assembly adopted Resolution A.981 (24), termed, 'New Legally Binding Instrument on Ship Recycling'. In accordance with the Resolution, the MEPC was required to develop a ship recycling system which provides regulations for:

- The design, construction, operation and preparation of ships, so as to facilitate safe and environmentally sound recycling, without compromising the safety and operational efficiency of ships;
- The operation of ship recycling facilities in a safe an environmentally sound manner and
- The establishment of an appropriate enforcement mechanism for ship recycling.

In addition to these regulations, the Resolution requires the MEPC to continue co-operating with the ILO and the appropriate bodies of the 1989 Basel Convention in this field, with the aim of avoiding duplication of work and overlapping of responsibilities and competencies among the three organisations. Furthermore, the Resolution urges governments and all involved stakeholders, in the meantime, to apply the IMO Guidelines on ship recycling without delay.

At its 54th session, the MEPC convened a working group on ship recycling facilities, which discussed the issues and further developed a draft text that includes articles and an annex with regulations for the safe and environmentally sound recycling of ships, including requirements for ships, ship recycling facilities and reporting. The MEPC also considered the report of the second session of the Joint International Labour Organization/International Maritime Organization/the 1989 Basel Convention Working Group on Ship Scrapping, which met in December 2005. The views of the group were taken into consideration by the MEPC Working Group on Ship Recycling.

At its 55th session in October 2006, the MEPC Working Group on Ship Recycling further developed the text of the draft Convention, providing globally applicable ship recycling regulations for international shipping and for recycling activities and it agreed to request the IMO Council, at its 98th session, to consider the provision of a five day conference in the 2008–2009 biennium, to adopt it.

Finally, the new Convention, 'The 2009 Hong Kong International Convention for the Safe and Environmentally Sound Recycling of Ships', was born at a diplomatic conference held in Hong Kong, China, from 11th to 15th May 2009, which was attended by delegates from 63 countries.

5.3 The Structure of the 2009 Hong Kong Convention

During the meetings of the IMO for the new ship recycling system, it has been discussed and agreed that, the system should comprise regulations for design (for new ships), construction, operation and preparation of ships (for new and existing ships), so as to facilitate safe and environmentally sound recycling but without compromising safety and operational efficiency. It is also important to consider regulations for the operation of ship recycling facilities in a safe and environmentally sound manner. Regulations for the establishment of an appropriate enforcement mechanism for ship recycling are also taken into account.

The 2009 Hong Kong Convention comprises 21 Articles, which cover the general obligations of Member States, definitions for the ship recycling process, application of the Convention, controls related to ship recycling, surveys and certification of ships, authorisation of ship recycling facilities, exchange of information amongst the Member States, inspection of ships which are subjected to recycling process, detection of violations, violations, undue delay or detention of ships, communication of information, technical assistance and co-operation, dispute settlement, relationship with international law and other international agreement enforcement proceedings.

The Articles are followed by the Annex to the Convention. The Annex comprises four chapters. Chapter 1 addresses the General Provisions of Regulations 1 to 3. Chapter 2 introduces the requirements for ships between Regulations 4 to 14. The requirements for ship recycling facilities are indicated between Regulations 15 to 23 under Chap. 3. Finally, the Chap. 4 notes the reporting requirements of Regulations 24–25.

In addition to the Annex, there are seven appendices established under the Convention. Appendix 1 deals with the control of hazardous materials. In order to control of hazardous materials, the appendix first lists the type of hazardous material, defines it and then notes the control measures. The 'minimum list' of items for the inventory of hazardous materials is introduced in Appendix 2. The Appendix 3 comprises the, *'form of the international certificate on inventory of hazardous materials', 'endorsement to extend the certificate if valid for less than five years where regulation 11.6 applies', 'endorsement where the renewal survey has been completed and regulation 11.7 applies', 'endorsement to extend the validity of the certificate until reaching the port of survey or for a period of grace where regulation 11.8 or 11.9 applies', 'endorsement for additional survey'.* In dealing with ship recycling facilities, Appendix 4 indicates the, *'form of the international ready for recycling certificate', 'endorsement to extend the validity of the certificate until reaching the port of the ship recycling facility for a period of grace where regulation 14.5 applies'.* Consequently, the form of the authorisation of ship recycling facilities is adopted under Appendix 5. Appendix 6 goes through the form of report of a planned start of ship recycling. Finally, Appendix 7 addresses the form of the statement of completion of ship recycling.

5.4 The Enforcement of the 2009 Hong Kong Convention

Article 3.1 of the 2009 Hong Kong Convention states that, unless otherwise expressly addressed in the Convention, the Convention shall apply to:

- Ships entitled to fly the flag of a Party or operating under its authority (3.1.1),
- Ship recycling facilities operating under the jurisdiction of a Party (3.1.2).

Article 3.1.1 allows the application of Convention in a broader range, in accordance with the term stated 'operating under its authority', rather than fully based on 'ships entitled to fly the flag of a Party'. However, this drafting also brings ambiguity, such as, what is the definition of, 'operating under its authority. The Convention does not define the operation of a ship, state's authority and how to limit it. This drafting is expected to lead to further problems in the future.

Despite this, it has been argued by some delegations during the discussions that the exclusions are not consistent with the spirit of Convention,[9] as Article 3.2 excludes the application of the Convention to any warships, naval auxiliary or other ships owned or operated by a Party and used for the time being, only on government non-commercial service. Article 3.3 states that, the Convention shall not apply to ships of less than 500 GT. Article 3.3 also states that, ships operating throughout their life only in waters subject to the sovereignty or jurisdiction of the State whose flag the ship is entitled to fly are excluded from the application of the Convention. For the non-party States, Article 3.4 states that the Parties shall apply the requirements of this Convention, to ensure that no more favourable treatment is given to such ships.

One of the most noteworthy points in the Convention is Article 16.4, which deals with States which comprise two or more territorial units in which different systems of law are applicable in relation to matter dealt with this Convention. This approach will allow States to apply the Convention to one of or two or more or all of their territorial units together or separately, if the State has two or more territorial units which have different law systems. This Convention aims to achieve as often as possible, signature, accession and ratification of States and in particular in

---

[9] Mikelis, note 3 above, pp. 1–10.

relation to China, as the one of the biggest participants in the ship recycling industry.

The entry into force of the conditions has been prescribed under Article 17 of the Convention and requires three conditions to be complied with simultaneously, for the enforcement of the Convention. These conditions are listed below:

- At least 15 or more States should sign the Convention, without reservation as to ratification,
- The combined merchant fleets of the States which have already signed the Convention should represent at least 40 or more percentage points of the gross tonnage of the global merchant shipping volume,
- The combined maximum annual ship recycling volume of the States which have already signed the Convention, should constitute at least three or more percentage points of the gross tonnage of the combined merchant shipping of the same States, during the preceding 10 years.

Article 15 regulates the relationship between the Convention and other international agreements. In accordance with Article 15.1, the Convention shall not prejudice the rights and obligations of any State under the United Nations Convention on the Law of the Sea (UNCLOS) and under the customary international law of the sea. In addition, Article 15.2 states that the Convention shall not prejudice the rights and obligations of Parties under other relevant and applicable international agreements. Even although the article aims to achieve co-existence of the international conventions and customary law with the 2009 Hong Kong Convention, such an article may cause several problems, such as:

- The inconsistencies in practice, amongst the international conventions, particularly UNCLOS and the 2009 Hong Kong Convention,
- The lack of global application of the 2009 Hong Kong Convention,
- The membership status of the States to (which are Parties to 2009 Hong Kong Convention) other international conventions.

5.5 The Impact of the 2009 Hong Kong Convention on the Ocean Law and Policy

The Current Capacity of World Merchant Marine Fleet In accordance with Article 17, the capacity of operating world merchant marine fleet is one of the key issues for the entry into force of the 2009 Hong Kong Convention. The world fleet changes continuously, old ships being replaced by new ones and whereas some of the old ships are being sent for recycling, some are being used for other purposes such as, stores, hotels or restaurants. In addition to that, some of the ships are removed from the world merchant fleet, as a result of an accident. Due, however, amongst other aspects to the non-existence of a global ship registration system, the flag of convenience and the lack of flag State's reporting and information, the certified capacity of merchant marine fleet varies from one database to another.

As Mikelis revealed in 2005, the capacity of the world merchant marine fleet over 100 GT comprises 92,105 ships and that over 500 GT comprises 47,258 ships.[10] By 2006, the capacity of the fleet over 100 GT had increased to 94,936 ships and that over 500 GT had increased to 49,213 ships. Sundelin states that the capacity of the merchant marine fleet over 500 GT was estimated as being more than 50,000 ships in 2008.[11] These reports, however, include ships which are owned by governments for non-commercial services or were engaged solely for domestic voyages. Article 3 of the 2009 Hong Kong Convention excludes ships which are owned by governments for non-commercial services and ships which are engaged solely for domestic voyages. Thus, the capacity of the merchant marine fleet which is subject to the 2009 Hong Kong Convention is estimated at about 42,000 to 45,000 ships.

Despite this, Mikelis states that the estimated request capacity for ship recycling are over 900 but less than 1,000 per year.[12] Vedeler reports that, each year, approximately 4,000 vessels are sent to recycling yards around the world.[13] A study by Andersen states that the annual expected scrapping rate is around 500–700 vessels per year.[14] The discrepancies amongst the reports are based on several reasons, amongst them being:

- Lack of reports and information for the ships which are subject to the recycling process,
- Contrary to first reason, while ships are already reported as recycled, some are subsequently found to have been traded onwards,
- Many of the ships that are subject to ship recycling have been reported a considerable time after they were recycled.
- Data reporting the details of ships subject to recycling varies from one year to another, varies from year to year due to lack of ship recycling reports on time and hence, the accuracy of the data is questionable,
- Differences in the reports published by the different maritime data providers,
- Inaccuracy of the ship recycling databases, particularly for the smaller sized merchant marine fleet.

---

[10] Nikos Mikelis, "A Statistical Overview of Ship Recycling," Proceeding of the International Symposium on Maritime Safety, Security and Environmental Protection, Athens, Greece, September 2007, pp.1–9.

[11] Sundelin, note 2 above, pp. 13.

[12] Mikelis, note 10, pp. 5–6.

[13] K. V. Vedeler, *From Cradle to Grave—Value Chain Responsibility in the Ship Scrapping Industry*, (Master's Degree Easy, Norwegian School of Economics and Business Administration, Norway, 2006), p. 32–33.

[14] Andersen AB, 'Worker safety in the ship-breaking industries. An issue paper' Geneva, ILO, 2001, available online: <http://www.ilo.org/public/english/dialogue/sector/papers/shpbreak/wp-167.pdf> last visited date: 20/10/2011.

Recently, in addition to the general factors of the ship recycling market, there are two main factors to consider. First of all, as a result of the global financial crisis, some ships have been laid up and then, the owners send them for recycling, in particular the old ships. Secondly, in order to try to prevent tanker incidents and marine pollution arising therefrom, the IMO set out a timetable for phasing-out single hull tankers. In accordance with the MARPOL Annex I, Regulation 13 G, the final phasing-out date for a Category 1 (pre-MARPOL) tanker was established in 2005. The final phasing-out date for Categories 2 and 3 tankers was first established in 2005 and then brought forward to 2015. The EU member States and some more leading maritime merchant States declared, however, that they will not allow the single hull tankers to sail through to their ports. This may be a reason for ship owners to renew their tanker fleets and therefore, send their aged single hull tankers for recycling.[15]

The Future Impacts and Prospect of the 2009 Hong Kong Convention on Ocean Law and Policy

The ship recycling industry has negative influences on the natural and the marine environment. On the other hand, the working conditions, occupational safety and health matters in recycling yards can pose a threat to the life and health of workers. Prior to the 2009 Hong Kong Convention being introduced, there were no specific internationally recognised standards addressing the above mentioned issues. The 2009 Hong Kong Convention will, certainly, have an impact on the designation of a global marine legal system and policy.

The authors will now discuss the possible future impact and prospects of the 2009 Hong Kong Convention. As is stated in the preamble, as per the 2009 Hong Kong Convention the Parties, concern is expressed about the environment, occupational safety and health, nonetheless, it is recognised that the ship recycling industry contributes to sustainable development. This drafting method aims to achieve a consensus between the economic demands and expressed concerns. Taking into consideration the experience gained from previous Conventions relating to the marine environment, the above mentioned concerns and the objective which the Convention would like to achieve, the 2009 Hong Kong Convention introduces the following control instruments:

- Flag State control system is established to ensure that the ships entitled to fly its flag or operate under its authority, shall comply with the requirements set in this Convention,
- The State in whose jurisdiction the ship recycling facilities operates, establishes a control system to ensure that the above mentioned ship recycling facilities comply with the 2009 Hong Kong Convention,

---

[15] Danish Environmental Protection Agency 'Shipbreaking on OECD' Working Report No. 17, 2003, available online: <http://www.mst.dk/homepage/default.asp?Sub=http://www.mst.dk/udgiv/publications/2003/87-7972-588-0/html/kap01_eng.htm> last visited date: 20/10/2011; Vedeler, note 13 above, pp. 34-35.

- The new surveying regime envisages an initial survey to verify the inventory of hazardous materials surveys during the operation life of the subject ship and a final survey prior to enter into the recycling process,
- The authorisation regime of the ship recycling facilities should be established by their States, in accordance with the 2009 Hong Kong Convention,
- The introduction of an information exchange system between the Parties,
- The list of hazardous materials whose installation or use is prohibited and/or restricted in ships, shipyards, ship repair yards or offshore terminals, has been prescribed under the 2009 Hong Kong Convention,
- The inventory of hazardous materials, specific to each ship, has been promulgated under the 2009 Hong Kong Convention,
- The ship recycling plan which comprises details relevant to subject ships, including its particulars and inventory,
- The introduction of an issuing process for an international, ready for recycling, certificate,
- The introduction of the ship recycling facility plan, which ensures attention is given to, among other matters, the occupational safety and health issues, appropriate information and training of workers, emergency preparedness and response plan, system for monitoring the performance of ship recycling and record-keeping systems.,
- The establishment and utilisation of prevention of adverse effects to human health and the environment procedures,
- The introduction of safe and environmentally sound management systems for hazardous materials,
- The establishment of reporting systems for incidents, accidents, occupational diseases and chronic effects,
- The introduction of a final reporting system, upon the completion of ship recycling.

These instruments aim to establish a comprehensive control and enforcement system for the effectiveness of the 2009 Hong Kong Convention, throughout the life cycle of a ship. From an environmental perspective, the 2009 Hong Kong Convention aims to create a comprehensive regime and to integrate it into the global marine environmental system. Article 1.1. states that, Party States are obliged to give full and complete effect to its provisions, in order to prevent, minimise and to a extent practicable, eliminate injuries and other adverse effects on human health and environment, caused by ship recycling activities. To achieve this comprehensive purpose, Annex Chap. 1, Regulation 3 stipulates that Party States shall take measures to implement the requirements of regulations of the Annex, taking into consideration the relevant and applicable technical standards, recommendations and guidance developed under the 1989 Basel Convention. In addition, Annex Chap. 2, Regulation 8 requires that ships which are subject to recycling shall: only be recycled at ship recycling facilities that are authorised in accordance with this Convention and fully authorised to undertake all the ship recycling process. As a result, ship recycling has been prohibited unless the ship

recycling process is conducted in a ship recycling facility, which is fully authorised. In addition, to avoid any adverse effects on the environment, the owners of
ships which are subject to ship recycling, shall conduct all operations in the period
prior to entering the ship recycling yard.

Article 4.1 sets forth the Flag State control system and to achieve this purpose,
Article 5 states that each Party State shall ensure that ships flying its flag or
operating under its authority, are subject to survey and certification. With regard to
these Articles, Annex Chap. 2, Regulation 10 stipulates that the Administration
(Flag State Administration) shall survey the ships by taking into consideration the
guidelines adopted by the IMO. Regarding to Regulation 10.4 and in every case,
the Administration shall be responsible in ensuring the completeness and efficiency of the survey and shall undertake to ensure the necessary arrangements to
satisfy this obligation

Article 6 of the 2009 Hong Kong Convention states that each Party State will
ensure that ship recycling yards will be operated under its jurisdiction. Furthermore, Article 4.2 states that each Party State shall require the ship recycling
facilities under its jurisdiction to comply with the requirements set forth in the
2009 Hong Kong Convention and shall take effective measures to ensure such
compliance. Detailed requirements for ship recycling facilities have been prescribed under Annex Chap. 3. Regulation 15 of the Annex states each Party States
shall establish legislation, regulations and standards, which are essential to ensure
that ship recycling facilities are designed, constructed and operated, in a safe and
environmentally sound manner. In accordance with Regulation 15.2, ship recycling facilities should meet the requirements of the 2009 Hong Kong Convention.
To achieve this purpose, each State Party is obliged to establish an authorising
mechanism for ship recycling facilities, whether or not those facilities meet the
requirements. In addition, Regulation 15.3 states that each Party State should
establish a mechanism for inspections of the ship recycling yards. The authorisation and inspection systems, which concern ship recycling yards, aim at an
integrated, comprehensive and continuous recycle controlling system.

Furthermore, Regulation 17 of Annex requires that ship recycling facilities shall
establish management systems, procedures and techniques, which should prevent,
reduce, minimise and where possible, eliminate adverse effects on the environment
caused by ship recycling, taking into consideration the guidelines adopted by the
IMO. In accordance with Regulation 17.2, ship recycling facilities are only
allowed to accept ships that:

- Comply with the 2009 Hong Kong Convention or
- Meet the requirements of this Convention.

In addition to these regulations, Regulation 18 of the Annex stipulates that ship
recycling facilities shall adopt a ship recycling facility plan and this plan shall
comprise;

- A policy which ensures environmental protection,
- An emergency preparedness and response plan,

- A system for monitoring the performance of ship recycling,
- A record-keeping system,
- A system which reports discharges, emissions, incidents and accidents causing damage or with the potential to cause damage, to the environment.

Ship recycling facilities shall ensure the safe and environmentally sound removal of any hazardous material contained in a ship, certified in accordance with Regulations 11 and 12 of the Annex. Waste management and disposal sites shall be identified, to provide for the further safe and environmentally sound management of materials. Furthermore, all waste which is generated from the recycling activity shall be kept separately from recyclable materials and equipment, labelled, stored in appropriate conditions and finally, transferred to a waste management facility which is authorised to deal with its treatment and disposal in a safe and environmentally sound manner.

For any emergency, the Regulation 21 states that a ship recycling facility shall establish and maintain an emergency preparedness and response plan. The plan shall ensure that information communication and a coordination system, which provides protection for all people and the environment is established and provides first-aid, medical assistance and fire-fighting services, plus, a strategy for evacuation and pollution prevention.

The 2009 Hong Kong Convention introduces a new system, which aims to exchange information between Party States via the IMO. If one of the Party States so requests, the State where the ship recycling facilities are authorised under its jurisdiction, shall provide the relevant information to the IMO. That information should be exchanged in a swift and timely manner. The 2009 Hong Kong Convention pays particular attention to inspections of the subject ships and states that the ships subject to recycling may be inspected by officers for the purpose of determining whether or not the ship complies with the requirements of the Convention. This inspection is mainly based on and limited to, a valid 'International Certificate on Inventory of Hazardous Materials' or 'International Ready for Recycling Certificate'. If, however, the ship does not carry a valid certificate or there are clear grounds for believing that:

- The condition of the ship or its equipment does not correspond substantially with the particulars of the certificate and/or the inventory of hazardous materials or

There is no procedure implemented on board the ship for the maintenance of an inventory of hazardous materials;

A detailed inspection may be carried, out taking into consideration the guidelines developed by the IMO.

In summary, the 2009 Hong Kong Convention establishes control and enforcement instruments in relation to ship recycling, including, Flag State obligations, Port State control rights, obligations of a Party and the recycling facilities under its jurisdiction, communication and exchange of information procedure, reporting system upon completion of recycling, detection of violations and auditing system.

Deficiencies Arising from the 2009 Hong Kong Convention

The 2009 Hong Kong Convention aims to achieve a comprehensive, globally and continuously applicable, environmentally sound manner ship recycling regime. Nonetheless, the structure of the 2009 Hong Kong Convention has deficiencies. Despite the entry into force of the conditions, the application of the Convention has experienced difficulties and this has lead to resultant problems. In addition, the 2009 Hong Kong Convention, itself, does not introduce a compulsory and environmentally ship recycling method.

The 2009 Hong Kong Convention establishes a comprehensive system, working on a timely basis, which covers instruments for such aspects as Flag State, Port State, State Part and ship recycling facilities under its jurisdiction, detection of violations, inspections and waste management. The system should take into account the periods before recycling, during recycling and after recycling. The Convention stipulates rules for the periods before the recycling and during the recycling only. Whilst the Convention stipulates the rules relating to waste management (after scrapping has been concluded), the waste management system does not address how to deal the management process during the final stage. The final management system of waste generated from the ships has not been stipulated in the Convention or has not been expressly integrated with the other international environmental protection regimes. Despite the Convention stipulating the co-existence of the relevant international legal instruments, in practice, problems may arise from such ambiguities.

Secondly, due to the lack of a global ship registration system, estimates of the future of the ship recycling industry are far from clear. As a result, the survey, inspection and reporting systems of the Convention may not work as well in practice, as is expected.

Thirdly, the application of the 2009 Hong Kong Convention does not comply with the ultimate aim of the environmental approach and has some significant exclusions. The new Convention excludes warships, naval auxiliary or other ships owned or operated by a Party and used only on government non-commercial service. Thus, the Convention is solely applicable to commercial ships. In addition, the Convention excludes ships which solely operate in waters subject to sovereignty or jurisdiction of the State whose flag the ship is entitled to fly. Finally, the new Convention is also not applicable for the ships of less than 500 GT. These exclusions limit the global application of the Convention and may thus limit its efficiency and success.

Furthermore, the 2009 Hong Kong Convention requires three pre-conditions and all these must be complied with simultaneously, before becoming effective. The first condition requires that at least 15 or more States sign the Convention, which is the easiest step to be achieved. As regards the signature, the EU Members and the OECD Members, have a consensus to sign and enforce the Convention. In order to this, the combined maritime merchant fleet capacity of these States must also constitute at least 40% of the gross tonnage of the world's merchant shipping capacity and consequently, meet the second pre-condition. Since, however, the third pre-condition depends on the signature of the five main ship recycling States, including Bangladesh, China, India, Pakistan and Turkey and among these, the only OECD Member State is Turkey, this poses a problem.

It is fortunate that China has the legal framework and preparedness order to meet the new Convention's requirements. By way of contrast, Bangladesh, India and Pakistan are still far from complying with the requirements of the new Convention and have reservations about signing the new Convention. Thus, the success of the 2009 Hong Kong Convention is questionable, at least, so far.

Last but not least, the new 2009 Hong Kong Convention is lacking in terms of stipulating compulsory and environmentally sound methods of ship recycling. There is no prohibition or restriction on grounding a ship/beaching. Even although the Convention requires pre-cleaning, survey and certification, grounding a ship on the beach may still cause marine environmental pollution. The grounding of a ship, in any way, offers potential risks to the marine environment and the new Convention does not introduce any improvement measures to tackle that matter. Based on the above discussion, the 2009 Hong Kong Convention is still far from achieving the ultimate aim of sustainable development.

5.6 Conclusion

The concept of environmentalism is an essential pillar of sustainable development. This concept has already influenced the IMO's attitude to the protection of the marine environment. The IMO has introduced conventions, regulations and established a global based marine environmental protection regime. At present, the IMO is concerned with the ship recycling industry and its potential adverse effects on the environment. Taking into consideration the economical issues, occupational safety and health issues and environmental issues, the IMO has introduced the 2009 Hong Kong International Convention for the Safe and Environmentally Sound Recycling of Ships, as a conclusion to the ad hoc meeting gathered in Hong Kong between 11th and 15th May, 2009.

The new 2009 Hong Kong Convention introduces a comprehensive regime to achieve environmentally sound ship recycling methods. The Convention obliges Party States to take appropriate measures and establish a domestic legal framework, in order to prevent, reduce, minimise and eliminate the adverse effects on the environment caused by ship recycling. The new Convention categorises the obligations of Port States and Party States, where the ship recycling facilities operate under their jurisdiction. Furthermore, the new Convention stipulates rules for the procedure of operation and recycling of ships, such as, survey and certification of ships, inspections of ships. In addition, the new Convention requires obligations of the ship recycling facilities, such as, preparing a ship recycling facility plan to exchange information if so requested and reporting upon the completion of the ship recycling. The general principles have been stipulated under the Convention and further details are explained and stated within the Annex and other relevant information, documents and lists are included under 7 Appendix of the Convention. This system shows that the IMO would like to establish a comprehensive and globally applicable ship recycling regime and integrate it with the marine environmental protection regime, by means of transition and the enforcement articles. By adopting the IMO's single hull phase-out regime and in recognition of the results of the global financial crisis, there are more ships than expected, to be sent for recycling. The above circumstances reiterate the significance and importance of, a global and comprehensive ship recycling regime.

Even although the new Convention is such an essential development for the marine environment protection regime and the industry, it still has deficiencies. These deficiencies arise from the structure of the Convention, due to the fact that it lacks of an integrated regime for the protection of the marine environment. Furthermore, the success of the Convention depends on the signatures of Bangladesh, China, India, Pakistan and Turkey. However, except for China and Turkey, the remainder have reservations about signing the Convention and this may affect its future viability. The application of the Convention is limited to ships of 500 GT or above thus, ocean-going commercial ships and so, numerically almost half of the ships sailing around the world are excluded from the Convention. In addition, the Convention does not offer a clearly defined method or methods for the recycling process and leaves this to the authority of the States' domestic law and regulations. The result is that the methods may vary from States. Even so, it is important to note that, the new 2009 Hong Kong Convention will offer a necessary improvement in the global marine environment protection regime, upon its coming into force.

Chapter 6
The Impact of Maritime Clusters

Abstract This chapter explores the meaning of the 'cluster' concept and how cluster theory has impacted on the maritime industry. This chapter also examines European and the United Kingdom perspectives of maritime clusters and concludes that there is a need to have a common definition of maritime sectors, clusters and a need to have consistent and updated data within Europe and the United Kingdom. The finding of this chapter is that both the European Union and the United Kingdom adopt a 'bottom-up' approach which in turn leading to a degree of fragmentation. A 'top-down' governance method instead should be introduced and this shift should be based on existing maritime clusters rather than creating new clusters.

Keywords Maritime clusters · European maritime policy · Economic benefits · 'Top-down' governance approach

6.1 Introduction

Although cluster theory is a relatively new concept, it has become the focal point of many new policy initiatives. This chapter explores the meaning of the 'cluster' concept and how cluster theory has impacted on the maritime industry. As observed, Europe is "the world's most competitive and dynamic knowledge based economy."[1] This chapter examines a European perspective of maritime clusters and concludes that there is a need to have a common definition of maritime sectors, clusters and a need to have consistent and updated data within Europe. Within the European Union, the United Kingdom's maritime clusters have the

---

[1] Christian Ketels, *European Clusters*, Hagbarth publication 2004, p. 1.

Y.-C. Chang, *Ocean Governance*, SpringerBriefs in Geography,
DOI: 10.1007/978-94-007-2762-5_6, © The Author(s) 2012

highest direct value added. Its approach to governance maritime clusters is, therefore, noteworthy. The finding of this chapter is that both the European Union and the United Kingdom adopt a 'bottom-up' approach. For the purpose of strengthening maritime clusters policy, a 'top-down' approach, instead, should be introduced.

6.2 Introduce the Cluster Theory to Maritime Industry

A cluster, has been generally understood as a group of similar things growing closely together.[2] One of the key features of cluster is that they share the same barriers in their external environment and can be removed only by joint action. Economists further defined the concept of cluster as "groups of companies and institutions co-located in a specific geographic region and linked by interdependencies in providing a related group of products and/or services."[3] Some critics argue that the definition of clusters is vague, thus, the concept itself is a problematic source of policy option.[4] Clusters exist, however, because firms or institutions find it beneficial to be physically close to other relevant firms or institutions which over time develop strong relations and interdependencies.[5]

The economic benefits may in turn stimulate the development of clusters theory. There are three reasons for developing clusters. Firstly, firms or institutions can operate with higher level of efficiency. This means that firms or institutions in clusters react quicker than they could in isolation. Secondly, firms or institutions in clusters working closely with customers and other companies create more new ideas and provide intense pressure to innovate. Since the cluster environment lowers the cost of experimenting, firms or institutions can, therefore, achieve

---

[2] P. W. de Langen, 'Clustering and Performance: the Case of Maritime Clustering in the Netherlands', Maritime policy & Management, 2002 Vol. 29, No. 3, p. 210.

[3] Christian Ketels, 'The Development of the Cluster Concept—present Experiences and Further Developments', paper prepared for NRW conference on clusters, Duisburg. Germany, 5th December 2003; See also Michael porter, *The Competitive Advantage of* Nations, (New York: The Free press, 1990); p. Doeringer and D. Tekla, 'Business Strategy and Cross-Industry Clusters', Economic Development Quarterly, Vol. 9, pp. 225–237; European Commission, *Regional Clusters in Europe,* Observatory of European SMEs, No. 3, Brussels 2002; paul Belleflamme, pierre picard and Jacques-François Thisse, 'An Economic Theory of Regional Clusters', (2000) Journal of Urban Economics 48, p 159; Christian Ketels, 'Cluster Development—picking or Energizing the League?', 2004, p 1, a research paper published on the website of the Institute for Strategy and Competitiveness, Harvard Business School, available from http://www.isc.hbs.edu/econ-clusters.htm last visited date: 13/08/2007.

[4] R. Martin and P. Sunley, 'Deconstructing Clusters: Chaotic Concept for policy panacea?', Journal of Economic Geography, 2003 Vol. 3 No. 1, pp. 5–35; European Commission, *Regional Clusters in Europe,* Observatory of European SMEs, No. 3, Brussels 2002.

[5] Gabriel R. G. Benito, Eivind Berger, Morten de la Forest, Jonas Shum, 'A Cluster Analysis of the Maritime Sector in Norway', International Journal of Transport Management 1 (2003), p. 205.

higher levels of innovation. Thirdly, the level of business formation tends to be higher in clusters and relies more on external suppliers and partners. The aforesaid circumstances will reduce the risk of failure, as entrepreneurs can rely on local employment opportunities in other companies in the same field.[6]

In addition, industries differ by the extent to which they can choose locations.[7] Many industries are tied to their location by the need to be close to their physical features,[8] for example, maritime industries are largely located around coastline areas. Clusters are particularly important for maritime manufacturing and other relevant services. Since most of the value-added products in maritime industries are produced by subcontractors, clusters can, therefore, offer subcontractors access to information and valuable knowledge that they cannot otherwise afford.[9] Through the participation of the general public and subcontractors, they will gain opportunities to develop expertise which may in turn lead to opening new markets, even outside of their own cluster.[10]

It is noteworthy that the development of maritime cluster needs to be based on existing manufacturing industries. For example, the unique position of ports within coastal areas and their role within the logistics chain of shipping and transport service have attracted special attention.[11] Although, originally ports were built to simply load and unload ships, they have now grown to become crucial industry and service hubs. In turn, port cities have become prime locations for the siting of industrial activities, for tourism and residential areas. Far from being dedicated to solely one activity, ports have now become truly multifunctional.[12] The development of transportation networks will further strengthen the effects of maritime industries to their hinterland. As a result of the aforesaid, a comprehensive decision making network such as maritime clusters needs to be employed in order to deal with the growing complexity of governance issues.

---

[6] Michael porter, 'Clusters and Competition: New Agendas for Companies, Governments, and Institutions', in *On Competition*, (Boston: Harvard Business School press, 1998).

[7] Christian Ketels, (2003), supra note 3, p. 5.

[8] These including markets (customers) and natural resources.

[9] Paul L. Weissenberg, 'European Economic Clusters: the European Commission perspective', pp. 18–19, in Niko Wijnolst ed., *Dynamic European Maritime Clusters*, (Maritime Forum, Norway and Dutch Maritime Network in cooperation with European Network of Maritime Clusters, September 2006).

[10] Ibid.

[11] Theo E. Notteboom and Jean-paul Rodrigue, 'port Regionalization: Towards a New phase in port Development', Maritime policy & Management, 2005 Vol. 32, No. 3, pp. 301–302.

[12] Dr. Joe Borg, 'The Future of ports—part of Our Future Vision for the Oceans and Sea', speech delivered on 28th June 2007.

6.3   European Perspective of Maritime Clusters

In Europe, maritime policy is traditionally organised on a sectoral basis.[13] Different types of regional maritime clusters have been developed in many parts of Europe. For example, a Belgium shipping cluster,[14] marine leisure[15] and sailing strategies[16] in Ireland, and a Dutch maritime cluster.[17] At present, there is no pan-European maritime cluster, and the effect of the European Union is, therefore, indirect. In order to reinforce cross-sectoral linkages between different sectors, a numbers of attempts have been initiated. The first was the Maritime Industries Forum (MIF) created in 1992 in order to facilitate the creation of synergies between maritime activities in various sectors and provide a permanent interface to the European Commission.[18] In 2003, the MIF established an Advisory Council for Waterborne Transport Research in Europe, subsequently named the Waterborne Technology platform, functioning as a forum where all stakeholders would have the opportunities to engage in a medium to long term vision.[19] The publication of the Waterborne Technology platform—*Vision 2020* and *Strategic Research Agenda*[20] formed a milestone of European maritime cooperation at the sectoral, national and European levels.[21] More recent development is the establishment of the European Network of Maritime Clusters (ENMC), which involve ten European countries in creating a platform for the exchange of best practices.[22] It is observed that there is a need to have a common definition of maritime sectors, clusters and a need to have consistent and updated data within Europe.[23] Based on the aforesaid, the ENMC has therefore recommended the Maritime policy Task

---

[13] Niko Wijnolst, 'Maritime Cluster policy and the Green paper', p. 127, in Niko Wijnolst ed., *Dynamic European Maritime Clusters*, (Maritime Forum, Norway and Dutch Maritime Network in cooperation with European Network of Maritime Clusters, September 2006).

[14] Policy Research Corporation N. V., *Economic Impact Study on Belgium's Shipping Cluster*, available from http://www.fisherassoc.co.uk/dbimgs/Belgian%20shipping.pdf last visited date: 20/08/2007.

[15] Marine Institute, Ireland, *Marine Leisure Strategy*, available from http://www.fisherassoc.co.uk/dbimgs/Ireland%20Strategy%20for%20Marine%20Leisure.pdf last visited date: 20/08/2007.

[16] Irish Sailing Association, *Strategic plan 2004-2008*, available from http://www.fisherassoc.co.uk/dbimgs/Irish%20Sailing%20Strat%20plan%20Summary.pdf last visited date: 20/08/2007.

[17] Presentation delivered by Drs. Ir. H.P.L.M. Janssens, at the Maritime Industries Firum, Bermen, 26th January 2005, available from http://www.fisherassoc.co.uk/dbimgs/Dutch%20Maritime%20Cluster.pdf last visited date: 20/08/2007.

[18] See http://www.marine.gov.uk/ec-mif.htm last visited date: 14/08/2007.

[19] Waterborne Tp, *Vision 2020*, February 2006, p. 5.

[20] Waterborne Tp, *Strategic Research Agenda*, May 2006.

[21] Niko Wijnolst, (2006), supra note 13, p. 128.

[22] European Network of Maritime Clusters, *Newsletter 2*, 2nd February 2006.

[23] Niko Wijnolst, (2006), supra note 13, p. 127.

Force to take action in order to fill the gaps.[24] The said initiative has a significant impact on maritime policy making. The result is, however, as yet to be seen.

"Strengthening maritime cluster awareness, not only at the national and European levels, but also at the regional and local level will contribute to enhance the maritime competitiveness of Europe as a whole."[25] A successful maritime cluster policy cannot be kept in an isolated, independent network. Instead, it requires efforts from many others. The current sector-based approach in Europe will be joined up by establishing a network between clusters that operate in the same or different sectors. It is believed that the aforesaid cooperation network will provide a platform for exchanging experiences, information, good practice and knowledge.[26] Consequently, a culture of 'outward looking' will be incorporated into the decision making framework, which will in turn lead to the development of joint research projects and business strategies. The MIF and the ENMC have already set up networks to bring about what is expected to be the end of the fragmentation of the maritime industries. Maritime clusters, therefore, would be one of the cornerstones of future European maritime policy.

6.4 The United Kingdom Approach in Relation to the Governance of Maritime Clusters

The United Kingdom's maritime clusters have existed for centuries. Until recently, however, different maritime sectors tended to work in isolation in terms of policy and political matters.[27] The said circumstances have led to a consequence that the United Kingdom maritime law, policy and other governance measures have traditionally been organised on a sectoral basis.[28] In the 1990s, strong bilateral relationships involving different maritime sectors emerged. Later, in recognising the value of different maritime sectors working together toward a better outcome, early in the 21st century a more multi-lateral approach began to emerge.[29] In recent years, the integration of marine environmental law and policy has been introduced as a new approach to the problem of fragmentation. This method can be appreciated in two senses, namely, to reflect external and internal changes. External integration addresses the problem of coordination across all ocean

---

[24] Ibid.

[25] Ibid, p. 120.

[26] Paul L. Weissenberg, (2006), supra note 9, p. 16.

[27] Mark Brownrigg, 'The United Kingdom's Maritime Cluster', p. 93, in Niko Wijnolst ed., *Dynamic European Maritime Clusters*, (Maritime Forum, Norway and Dutch Maritime Network in cooperation with European Network of Maritime Clusters, September 2006).

[28] Yen-Chiang Chang, *Good Governance in the Management of Marine Environment—A Comparison of Taiwan and Scotland*, phD thesis September 2007, School of Law, University of Dundee, pp. 170–187.

[29] Mark Brownrigg, (2006), supra note 27, p. 93.

governance public authorities by promoting the consideration of marine environmental issues across all policy areas. Internal integration addresses the problem of
fragmentation amongst diverse regulatory agencies by drawing together regulatory
responsibilities for different environmental emissions and impacts within one
single authority.[30] This is, however, still in evolution both deepening and
spreading.

In the United Kingdom, there are three official organisations which deal with
marine affairs. As detailed by Hance D. Smith and Jonathan S. potts, they are, the
Department for Environment, Food and Rural Affairs (DEFRA), the Department of
Trade and Industry (DTI) (is now Department for Business, Enterprise and Regulatory Reform, DBERR)[31] and the Department for Transport (DT).[32] Among
them, DBERR is specifically responsible for cluster policy.[33] Much of the
responsibilities of DSERR, however, have been transferred to regional authorities
under the Regional Development Agencies Act 1998.[34] DSERR's approach to
regional policy is designed to build the capability of regions, putting greater
emphasis on growth within all regions and strengthening the building blocks for
success and boosting regional capacity for innovation and enterprise.[35] Based on
the aforementioned, it can be concluded that the current United Kingdom approach
to cluster activities in the maritime sector is operated both sectoral and also multi-
tiered, at national, regional, sub-regional or local level. As observed by Mark
Brownrigg, the United Kingdom maritime clusters are "built very much on
continuing dynamic activity by a wide range of organisations across a very diverse
maritime economy."[36]

The commercial orientation of the United Kingdom maritime clusters also finds
its expression in different ways and at different levels. In general, commercial
activities tend to be either at regional or local level rather than national level. As a
result, many such initiatives have attracted government and EU funding. The most
prominent examples of commercial maritime clusters can be found from Maritime
London, Marine South West, Mersey Maritime, Marine South East.[37] Other
groupings also exist, such as Maritime plymouth at local sub-regional level.
Marine South West together with Maritime plymouth has made this area attractive

---

[30] Stuart Bell and Donald McGillivray, *Environmental Law* 6th ed. (Oxford: Oxford University
press, 2005), p 107; John F. McEldowney, *public Law*, (Sweet & Maxwell's Textbook Series,
2002), p. 278.

[31] DBERR brings together functions from the former Department of Trade and Industry,
including responsibilities for productivity, business relations, energy, competition and consumers,
with the better regulation executive, previously part of the Cabinet Office.

[32] Hance D. Smith and Jonathan S. potts (ed), *Managing Britain's Marine and Coastal
Environment—Towards A Sustainable Future*, (London: Routledge, 2005), p 84.

[33] European Cluster Observatory, *Cluster policy Report—United Kingdom*, June 2007, p. 2.

[34] Section 6 of the Regional Development Agencies Act 1998, Chapter 45.

[35] European Cluster Observatory, (2007), supra note 33, p. 2.

[36] Mark Brownrigg, (2006), supra note 27, p. 93.

[37] Ibid, pp. 98–99.

as a major base for marine leisure and events, and for marine and maritime research, science and technology, including centres of expertise.[38] It is not difficult to discern that in the United Kingdom a 'bottom-up' governance approach is adopted to deal with maritime clusters issues.

There is an attempt towards a more 'top-down' approach in the United Kingdom. For example, the latest development of maritime clusters activities is co-ordinated by the Sea Vision UK, which intends to combine the commercial clusters at national level, with regional inputs on policy issues.[39] It provides a platform for different national sectors and regional clusters to exchange information on relevant developments and take joint action where that is considered to be appropriate. This initiative is, however, still in its infancy. The Sea Vision UK's attempt to coordinate different sectors and levels remains subject to further observation.

6.5   Conclusion

Given a broad picture of the development of maritime clusters, it is not difficult to conclude that a 'bottom-up' approach is adopted at both European and the United Kingdom levels. Although it is highly desirable for a competitive maritime cluster to have foreign partners, most policy initiatives address regional maritime clusters and actors within their administrative jurisdictions. To this end, this chapter suggests the introduction of a 'top-down' governance approach. This means supporting existing maritime clusters and helping them to bridge the gaps between different sectors would perhaps be a better policy option than creating a new cluster. The focus is then on providing a platform for dialogue and cooperation between the general public, higher education, research institutions and private organisations at local, national, European and international levels. By this means, a sound maritime cluster must be able to identify market failure, provide cluster governance and advice, develop links between industry, research institutions and universities in order to offer appropriate education and training courses, create hubs for networking and exchange of information, and finally provide suitable infrastructure and financial support.[40] The development of maritime clusters should not aim to create new clusters but to activate them.

---

[38] Fisher Associates, *plymouth Marine Sector Development Strategy*, June 2001, p. 2.

[39] See http://www.seavisionuk.org/what_is_sea_vision.htm last visited date:15/08/2007.

[40] Paul L. Weissenberg, (2006), supra note 9, p. 21.

Chapter 7
Can the Social and Cultural Impacts of Ports be Assessed in Terms of Economic Value?

Abstract This chapter firstly recognises the importance of the port heritage and suggests the economic value of ports can be assessable only if the port heritage is restored to 'economics'. There would not be an economic problem if the port heritage is treated as ethically oriented. This chapter then emphasises that the historical heritage element of port is considered to be a common potential to develop tourist activities. The tourist-historic pattern may provide an incentive for other developments which in turn may boost the local economy. As a result, social and cultural impacts of ports can be assessed in terms of economic value through the tourist-historic pattern.

Keywords Maritime heritage · Economic value · Maritime museum · Port heritage · Geographical location · Tourist-historic pattern

7.1 Introduction

The scope of social and cultural sectors of ports is broad and therefore vague in content. In order to consider this issue further, it might be appropriate to start by asking the question 'What does the ocean mean to the United Kingdom?'. Everything, in the briefest possible desired to the question. "The United Kingdom is the greatest maritime nation in the world," a significant amount of people who living in this island will answer this question in a careless, non-understanding way.[1] It is observed that although there is plenty of sentimental interest in the sea, there is a lamentably small amount of practical knowledge of these great matters.[2] It is undoubtedly true that in considering what the sea means to us, marine

---

[1] Frank T. Bullen, *Our Heritage the Sea* (London: John Murray, Albemarle Street, 1906), p. 303.
[2] Ibid.

resources, seaborne trade and other human activities will come to most people's mind.[3] Along with the commercialisation and industrialisation of the global economy, sea-related activities have historically been organised and governed in a separate and distinctive way. Thus, seafarers such as fishing villages, seaborne trade and naval bases are still separate communities.

The aforesaid circumstances will derive a consequence that the sea-related activities were relatively isolated from the land and inevitably leading to differences in traditions and outlook.[4] More specialised maritime groups have emerged in the current maritime development, such as the leisure industries, offshore oil and gas, marine science, education and marine conservation.[5] All of the aforementioned have bequeathed a strong maritime heritage and tradition in the United Kingdom and elsewhere. This chapter firstly recognises the importance of the port heritage and suggests the economic value of ports can be assessable only if the port heritage is restored to 'economics'. This chapter then emphasises that the historical heritage element of port is considered to be a common potential to develop tourist activities. The tourist-historic pattern may provide an incentive for other developments which in turn may boost the local economy. As a result, social and cultural impacts of ports can be assessed in terms of economic value through the tourist-historic pattern.

7.2 The Maritime Heritage of Ports

Maritime heritage differs which has emerged from the long sequence of the development path. Navigation and shipping can be traced back to the beginning.[6] Ports, functioning as cargo transfer,[7] did not become meaningful until Roman times.[8] Shipping together with ports serves as a pivotal use of the sea in various ways. Early stages of the ports' heritage development focus on the realms of heritage conservation projects such as the oldest docks, custom houses and warehouses.[9] At a later stage, attention had been paid to the large-scale urban

---

[3] Ibid, p. 305.

[4] Hance D. Smith and Jonathan S. Potts, 'People of the Sea', in Hance D. Smith and Jonathan S. Potts ed., *Managing Britain's Marine and Coastal Environment—Toward a Sustainable Future*, (London and New York: Routledge, 2005), p. 7.

[5] Ibid.

[6] Ibid, p. 11.

[7] James Bird, *Seaports and Seaport Terminals* (London: Hutchinson, 1971), p. 74.

[8] Sarah Jane Tuck, *Socio-Economic Aspects of Commercial Ports and Wharves in Southwest England: A Grounded Theory Approach to Regional Competitiveness*, Doctor of Philosophy Thesis, Business School, Faculty of Social Science & Business, University of Plymouth, July 2007, pp. 27–28; see also Hance D. Smith and Jonathan S. Potts, supra note 4, p. 11.

[9] James Bird, supra note 7, pp. 66-72.

renewal projects such as London Docklands, the Mersey Docks and Cardiff Bay[10] all of which the port function is largely reduced and replaced by industrial and residential land uses. The ports also provide the primary location for the maritime museums.[11] For example, the National Maritime Museum at Greenwich, with other notable examples in Liverpool, Newcastle upon Tyne, Aberdeen and others. Another important adjunct to the maritime museum is the nation's collection of historic ships such as the *Cutty Sark* at Greenwich[12] and the *Discovery* at Dundee.[13] The aforementioned port heritage will in turn generate immense educational and recreational values.

The naval port is another aspect that needs to be addressed. Naval ports have been designed for the defence of the realm and therefore normally have been closely linked with the fortunes of associated port towns. This is mainly due to the need for naval use of maritime space will restrict the local commercial port development.[14] Often, naval ports are geographically quite separate from commercial ports and concentrated in such locations with strategic significance such as: Pembroke Dock, Plymouth, Portsmouth, Chatham, Deptford and Rosyth.[15] Consequentially, the aforesaid naval ports have stood outside the commercial port system and thus suppressed market economies.[16] It is, however, worth noting that there is an increasing interaction of geopolitical, technological and economic controls over naval ports. This may in turn force governments to economise and induce a shift of emphasis away from the navy in many naval port cities.[17]

To this end, it is important to note that "local policy-making on the historic environment takes proper account of the value a community places on particular aspects of its immediate environment."[18] The character assessment to local authority, therefore, is a useful tool as a way of encouraging greater local involvement in port heritage issues.[19] To consider the aspect of the historic environment in promoting economic, employment and educational opportunities within the local scale is essential when preparing community development strategies.[20]

---

[10] Ibid.

[11] Hance D. Smith and Jonathan S. Potts, supra note 4, p. 12.

[12] See http://en.wikipedia.org/wiki/Cutty_Sark last visited date: 04/09/2007.

[13] See http://www.dundeecity.gov.uk/photodb/wc0805.htm last visited date: 04/09/2007.

[14] Brian Hoyle, Philip Wright, 'Towards the Evaluation of Naval Waterfront revitalisation: Comparative Experiences in Chatham, Plymouth and Portsmouth, UK', Ocean & Coastal Management 42 (1999), pp. 957–958.

[15] Hance D. Smith and Jonathan S. Potts, supra note 4, p. 13.

[16] Brian Hoyle, Philip Wright, supra note14, pp. 957–958.

[17] National Audit Office, *Ministry of Defence: Transfer of the Royal Dockyards to Commercial Management*, 3rd March 1988, pp 11–17.

[18] Department of Culture, Media and Sport, *The Historic Environment: A Force for Our Future*, (2001), para. 3.16.

[19] Ibid, para. 3.19.

[20] University of Plymouth, *Noss-on-Dart Feasibility Study* (2003), para. 6.11.

7.3 The Economic Value of Port Heritage

The oceans have different kinds of economic values, some of which are very difficult or impossible to be quantified. Some marine resources can be quantified and managed in monetary terms, for example, the form of individual transferable quotas has been implemented to manage the fisheries resources. This form of fisheries privatisation means that individual fishers or fishing companies are allocated 'quotas' by the fishery authority. The said quotas may in many ways become fishermen or fishing companies' private property. Based on ownership, fishermen or fishing companies have the rights to exploit certain amount of fisheries resources at their convenience throughout the seasons. This ownership is tradable which means that fishermen or fishing companies are able to sell their quotas or licences,[21] although this is subject to fisheries legislation. Under the aforesaid circumstances, economic value can be imposed on the fisheries resources. Can the same approach apply to the port heritage? In other words, can port heritage be quantified and expressed in monetary terms?

There are two prerequisites that need to be considered in order to discuss the economic value of port heritage. Firstly, it is important to recognise that the port heritage exists, which has been discussed earlier in this chapter. Secondly, there would not be an economic problem if the port heritage is treated as ethically oriented. This issue will be meaningful only if port heritage is restored to 'economics'. In addressing economic value of port heritage, the alternative is strongly driven towards the latter. Although classical economics comprises only what can be quantified and expressed in terms of pounds or dollars, this view will limit and distort the real wealth of people, of nations and of the world.[22] For example, port heritage comprises social, cultural, geographical and ethical values that are difficult to be quantified and forced into the market system. It can be concluded that when evaluating the economic value of port heritage a certain degree of uncertainty will exist.

Uncertainty may be caused by lack of information and reliable data which may in turn generate risk. The greater the uncertainty, the higher the risk. The degree of risk differs from port to port, as each has its unique regional features and network. The success of the development of port heritage is seen as being very much related to geographical location. Location is relevant to the ports' ability to attract investment, tourists and be in a position to develop commercial port activity. It can be observed that different ports are capable of attracting different types of new business or economic activity. As an official from the Plymouth port authority states that

---

[21] E. Mann Borgese, 'The Economics of the Common Heritage', Ocean & Coastal Management 43 (2000), p. 767.

[22] Ibid, p. 771.

I do not feel that waterfront redevelopment involving containers or roll-on/roll-off is possible; it is very difficult to see that we could increase trade because of our geographical location, and even if the dockyard could be used for commercial shipping, it is very difficult to gain access to it, and the only part of the dockyard which could be beneficial to commercial activity, commercial shipping, would be the north end where there is deep water, but of course that is the part the navy would want to retain.[23]

This point is further strengthened by a statement made by a Ministry of Defence representative

Geographical location plays an important part in successful redevelopment. Portsmouth, when looked at in this light, is a straightforward journey from London; Plymouth, whilst it might be straightforward, takes between four and five hours, and Chatham—well, historically it is not regarded as in the main stream of things, and it has got a lot to overcome. Of course three which we generally consider, Portsmouth is rather better placed than the others.[24]

Although the above examples might not be representative, they demonstrate that the geographical location of the ports combined with their local, regional and historical attributes form a component part of port heritage. All of the above-mentioned indicators may in turn determine a port's ability to benefit from any land release and to develop an appropriate programme.

The historical heritage element of a port has immense value for education purposes as, "the fabric of the past constitutes a vast reservoir of knowledge and learning opportunities."[25] Based on this, ports heritage is considered to be a common potential to develop tourism.[26] The exploitation of the port heritage along with tourist-historic lines may provide an incentive for other developments. Together with the historic buildings, custom houses, warehouse, dockyard, famous maritime events, personalities and ships which will provide a considerable resource for the generation of 'maritime heritage image' and in turn prove influential for external investors. For example, in 1996 Bristol hosted an *International Festival of the Sea* which was deemed to be the largest maritime festival ever staged in the United Kingdom. The event involved the reorganisation of the material spaces of the host city and promoted Bristol as a 'maritime heritage' landscape.[27] The long term intention was to attract national and international investment by 'place-marketing' the city as a distinctive location boasting plentiful character and history.[28] The implications of these developments are enormous. "Economically motivated cultural restructuring involves not only a revision of the

---

[23] Citing from Brian Hoyle, Philip Wright, supra note 14, p. 976.

[24] Ibid, p. 979.

[25] Supra note 20, para. 6.5.

[26] Ibid, para, 6.6.

[27] David, Atkinson and Eric Laurier, 'A Sanitised City? Social Exclusion at Bristol's 1996 International Festival of the Sea', Geoforum, Vol. 29, No. 2, 1998, p. 199.

[28] Financial Times (1994) Bristol, *Financial Times* 29th November 1994, I–VIII.

past, but a deployment in the conflictual space of the society being 'touristified' of desiring and polemical images of identity and alterity in the future."[29]

It can be predicted that marketing and subsequent management are critical parts of evaluating the economic value of port heritage. The shift in emphasis from the port heritage to the tourist attraction is seen as allowing a variety of new business and local taxpayers to contribute to the local fiscal base and thus strengthen the local economy.[30] Overall, it would be appear that port heritage is perceived as beneficial to the local community. Diversification is seen as an inherent component of port heritage so that the port city does not have to rely on one major employer. There is a need to market and manage port heritage in order to ensure the success of this scheme. In an environmental connotation covering both human actions and environmental consequences, it comprises both "a series of technical and generally well defined measures governing physical interactions between human activities and marine environment; and [...] a general management dimension encompassing co-ordination of technical management measures, organisational decision-making, policy and strategic planning aspects."[31]

7.4 Conclusion

Can the social and cultural impacts of ports be assessed in terms of economic value? Yes, they can, although this is difficult. This chapter suggests that social, cultural, geographical and ethical values are component parts of the port heritage which are difficult to quantify and be encapsulated into the market system. In order to evaluate the economic value of ports, port heritage has to be restored to 'economics'. This chapter also suggests that tourist-historic pattern is one of the approaches which can encourage developers or investors to explore the potential business of port heritage. Since the port heritage has a strong link to local character, the development of the port heritage will in turn strengthen the local economy. Marketing and managing port heritage is a way to ensure the success of tourist-historic scheme. Port heritage is a record of the human use of the sea in all its diversity over a range of time scales.[32] Furthermore, 'maritime heritage image' itself has become a major economic resource, not only in what remains and continues to develop in specialist maritime communities, but also in the museums, archives, restoration and elsewhere which forms the basis of the educational and recreational purposes.

---

[29] M. Morris, 'Life as a Tourist Object in Australia', in M. F. Lanfant, J. B. Allcock and E. M. Bruner ed., *International Tourism, Identity and Change* (London: Sage, 1995), p. 183.

[30] Brian Hoyle, Philip Wright, supra note14, p. 972.

[31] Hance D. Smith and Jonathan S. Potts, supra note 4, p. 14.

[32] Hance D. Smith and Jonathan S. Potts, supra note 4, p. 17.

Chapter 8
Port Governance in the South West of England: A Comparative Assessment

Abstract The South West of England is a very extensive region with constraints in terms of its transport network. It is imperative if the region is to optimise its economic development for transport services to fully utilise all transport modes, not least short sea shipping. To achieve this utilisation will require, inter alia, investment in the region's port infrastructure. In addition, the region is well-placed geographically in relation to a number of other European Union countries to develop seaborne trade links with them. If this can be supported both by the development of coastal shipping links between regional ports and the incremental capability of rail and road transport within the region then substantial increased economic activity for the region could result. Moreover, it will be beneficial if the region's ports could act as a network, optimising the contribution each port can make. This in no way precludes healthy competition between the ports.

Keywords Port governance · The South West of England · SWOT analysis · Comparative assessment

8.1 Introduction

In the South West of England, with the peninsula having both a north and south coast, there is an intimate connection between the sea and land. This combined with a temperate climate and relatively unspoilt countryside and coastline, has not only fostered a distinctive maritime industry but also made the South West of England an important tourist destination. On the other hand, transport networks particular road and rail connections are considered poor and thus hinder connections with the rest of the UK and Europe. For example, one major truck road is hampered by twist traffic in the summer and there is only one rail link which is affected by storms on the south coast between Exeter and Plymouth. In both cases, the north part of the region is very poorly served. This in turn has undoubtedly

Y.-C. Chang, *Ocean Governance*, SpringerBriefs in Geography, 117
DOI: 10.1007/978-94-007-2762-5_8, © The Author(s) 2012

contributed to a survival mentality as opposed to a growth mentality and has hampered the development, of many maritime activities.

It is the intimate relationship between maritime activities and tourism that gives the South West Marine and Maritime Sector its distinctive flavour, thereby helping to distinguish it from marine sectors in other regions. In order to provide an comparative assessment, port profiles including port ownership and port facilities have been developed enabling an overview and general comparison of port characteristics. The following has been gathered via interview and general existing information on individual ports available. Key policy driver and operational objectives for governing the ports in the South West of England are provided after a discussion on SWOT analysis.

8.2 Port Ownership

There are significant differences in the nature of port ownership—UK port ownership is diverse. However there are three main types in which UK ports can be categorised which is shown by table one. The majority of the ports in the South West of England are trust or municipal ports rather than private/commercial ports. In fact, through secondary research, the author found that there were inconsistencies in port type. A reason for this is that although ports have been categorised as either 'municipal', 'trust' or 'private', with some ports this may depend on the area. For example, Falmouth is a trust port, however the dock area is privately controlled. Another example is Plymouth port—the four specific areas which comprise Plymouth port is a mixture of naval, trust and private port ownership and management. It must be assumed in this case that the table one refers to the *main ownership* of port areas.

8.3 Port Facilities

Based on the information collected for the port profiles table two gives an overview of basic port information which can be summarised and were most consistently and readily available from secondary sources. This includes vessel length (metres), maximum draught (metres) and berth information according to port.

In terms of vessel length and maximum draught the port of Bristol provides the largest capacity (300 m). Falmouth also provides large capacity for tankers (240 m) and also Medway port/Sheerness provides large vessel length capacity at 230 m length. Truro provides a large lay-up berth at 219 m, and Plymouth also provides 200 m maximum length at their Millbay Docks area. Poole and Fowey all provide vessel length capacity between 150 and 175 m (160, 150 respectively). Ports which provide vessel length between 119 and 130 m are Weymouth, Par and Teignmouth (130, 125, 119 respectively), with the much smaller ports and harbours—Torbay much lower vessel length capacity. In terms of maximum draft, the majority of ports appear to have areas which accommodate 8.5 m—specifically Fowey, Plymouth, and Falmouth providing 8.4 m Smaller maximum draft capacities is Truro (4.4 m).

Smaller ports and harbours for example those in Torbay Harbour generally accommodate more fishing industry (primarily Brixham) and leisure craft oriented activity, therefore vessel length, maximum draft and berthing information would not be a particularly relevant comparison to the larger commercial ports. These are just some of the factors which define UK ports as having differing attributes.

It is important to remember that this initial sweep of information does not include the context of all specific berth length and number in the first instance, and also the comparison has been made on specific sections of the port *where the data was available*. It is possible therefore that some ports in the table have been considered as a whole, and some have been separated into sections and this is purely due to information which was available at the time of the preliminary secondary research.

8.4 Discussion: SWOT Analysis

As a result of the analysis and consultation process undertaken during the study a number of issues arose which reflect and also shape the ports sector in the South West of England, and these are outlined below.

Strengths

- The ports are generally well equipped with modern, efficient infrastructure and flexible working conditions and provide a high level of service, at competitive rates, to port users.
- The ports have kept pace with changes in the market place to the benefit of port users. A strong example is the introduction of new technology such as high speed ferry services offering reduced journey times and increased frequency of services.
- The ports operate within an intensively competitive environment, between South West of England ports and key South East of England ports. This environment provides competition and choice in the provision of port services to the region with consequent user benefits including lower port charges and an enhanced network and frequency of shipping routes.
- The South West of England's geography allows a range of services to be offered (e.g. short sea, overnight etc.) to suit customer preferences.

Weaknesses

- Transport to/from ports is mainly by road. Certain ports are rail fed, including Bristol, Par, Poole and Fowey.
- There is a concentration of port activity to the benefit of Plymouth and Bristol, which reflects trends towards rationalisation of the ports sector experienced at both national and European levels.

- There is a shift within the shipping sector towards use of larger sized vessels for transporting an increasing range of goods which means the lack of deep water places some facilities at a competitive disadvantage (e.g. Par).
- Environmental designations at certain ports may constrain future development.

Opportunities

- Tourism is expected to continue to increase, for example The Eden project, with the foreseen increase in short breaks and overseas tourist visits, and the development of niche markets. The ports of the South West of England should benefit from increasing numbers of tourists travelling by sea (e.g. cruise liners, visiting pleasure craft).
- The network of ports along the south coast of the South West of England offers an ideal opportunity for coastal shipping for certain trades.
- Short sea shipping within a European network opportunities exist at some of the region's key ports, and these opportunities could be exploited.

Threats

- The ports of the South East of England and South Wales continue to strengthen as a result of significant infrastructure investment, the adoption of more flexible labour arrangements and a commitment to lower charges. This threat is faced by both freight and passenger trades, and investment to maintain existing infrastructure is a priority.
- New technology and modern infrastructure is continually changing and the ports will need to adopt an ongoing programme of infrastructure improvement/ upgrade in order to maintain current high standards and competitive environment within a European context.

8.5 Key Policy Driver

A key component of the ports strategy is the potential role of short sea and coastal shipping in reinvigorating a number of the region's gateway ports and promoting a sustainable method of transporting goods in and out of the region. In the context of this strategy, in broad terms short sea shipping is defined as shipping services between the UK and Ireland/Continental Europe, whilst coastal shipping refers to shipping services (typically in vessels of under 4,000 dead weight tonnage) between ports along the UK coastline and inland waterways. The author has identified two key factors shaping the development of coastal shipping in the South West of England:

- The likely target hinterland for coastal shipping will be away from those hinterlands readily accessible by road transport, such as Bristol and Poole.
- The role of Plymouth as both a source and destination for coastal shipping should grow.

Consequently, this defines the potential hinterland and area for coastal shipping to the west of a line from Exmouth to Barnstaple. Also, the number of potential ports suitable for coastal shipping is limited by the cost of investment and length of sea passage. Thus, it is recommended that Plymouth and one other port should be promoted as a hub for coastal shipping services in the South West of England. The north coast harbours have disadvantages in size, access and passage time, thus it is also suggested that goods would naturally move from Plymouth to and from the north coast. Consideration was given to both Fowey and Falmouth as an alternative hub for coastal shipping, however the former has restricted land side operational facilities and the latter is less centrally located in relation to Cornwall's markets.

It is anticipated that the development of the South West of England economy will generate exports of minerals or manufactured goods and the import of raw materials or manufactured goods. The author briefly touches on each trade in turn:

- Export of minerals—Short sea and coastal trade from South West of England ports currently exists in clay and stone. The coastal shipping of dressed and non dressed hard stone plus crushed stone could well increase given the lack of such minerals in the South of England. Transport costs are of the essence with such commodities and dictate either rail or sea transportation.
- Export of manufactured goods—The decline of manufacturing in the South West of England has reduced volumes. If quantities are small and speed is important then time taken to consolidate loads is a detriment to competitiveness. Large quantities and low load consolidation times could produce possibilities for coastal shipping.
- Import of raw materials—Speed may be less important and if the industrial base is being developed then it may be possible to push the trade in a particular direction via grants/concessions etc.
- Imports of manufactured goods—Except for non time sensitive unit loads this is unlikely. Break of bulk/destuffing at entry is more likely, though with "Channel loops" it may be possible, since such activity is EU inspired.

Coastal shipping is a growing European transport mode. The shape and geography of the South West of England is well suited to the advantages of coastal shipping both for "distribution/intra regional" cargoes and for wider trade with the rest of the UK. The sustainable nature of this mode is of key importance in a region with an overburdened infrastructure and should form an important component of the ports strategy. It is an attractive alternate transport mode for the region and will contribute to port development. However, its applicability to industries developing in the South West of England and the necessary changes to transport practices require detailed study.

The promotion of short sea shipping and waterborne transport in general, is now a policy priority at European level of both the European Commission and of the Member States. The principal motivator for this priority is the environmental benefits of short sea shipping and waterborne transport compared to other transport modes. The environmental disbenefits of road transport in particular have led to a promotion of waterborne transport as an alternative means of relieving the increasing road congestion and environmental damage caused by further expansion of trucking as a means of non-bulk freight carriage.

8.6 Key Operational Objectives

The region's ports are important drivers of economic growth. Through their role as regional gateways they are able to stimulate a whole chain of value added activities within the regional economy through developing supplier linkages, associated business parks, attracting inward investment, stimulating tourism etc. The key challenge of the Regional Gateways Strategy is to encourage initiatives that strengthen these economic linkages and maximise local and regional economic development. In this context, the key components of the Regional Ports Strategy are:

- Support the improvement of land based links to the region's ports, with the emphasis on the most sustainable means of transport.
- Support the development of each port in its individual roles where such development provides economic benefit and can occur without environmental damage.
- Encourage the promotion of coastal and short sea shipping opportunities at selected gateway ports.
- Support development of life line services to the peripheral parts of the region.
- Support tourism and economic regeneration role of ports through water based leisure activities, cruise shipping and visiting pleasure craft.
- Support co-located commercial development at gateway locations, with a particular emphasis on strategic sites as identified by South West Regional Development Agency.

8.7 Conclusion

The South West of England is a very extensive region with constraints in terms of its transport network in relation to Devon and, particularly, Cornwall. It is imperative if the region is to optimise its economic development for transport services to fully utilise all transport modes, not least short sea shipping. To achieve this utilisation will require, inter alia, investment in the region's port infrastructure. Moreover, it will be beneficial if the region's ports could act as a network, optimising the contribution each port can make. This in no way precludes healthy competition between the ports. European experience shows that co-operation and competition are not incompatible. It is important in the South West of England to appreciate that this same principle applies to the various transport modes.

The development of short sea shipping or rail does not jeopardise the future of road transport, rather it enables the constraints imposed by road congestion to be overcome and for each mode to play its optimum role in a multi-modal transport system geared to the needs of the manufacturing and distribution companies whose goods need to be transported within the regions, the UK, and Europe. Moreover, the development of one mode is likely to require investment in other modes. It is crucial to recognise and assess the overall logistics requirements which the transport modes in combination are established to meet.

The South West of England region is well-placed geographically in relation to a number of other European Union countries (i.e. France, Ireland, Portugal and Spain) to develop seaborne trade links with them. If this can be supported both by the development of coastal shipping links between regional ports and the incremental capability of rail and road transport within the region then substantial increased economic activity for the region could result. It is also the case that the application of information and communications technology (ICT) will be required to achieve the optimisation of the region's transport network, since the utilisation of ICT may improve the overall business performance. This is also with potentially significant spin-off benefits for the ICT sector in the region.

Index

Y.-C. Chang, *Ocean Governance*, SpringerBriefs in Geography,
DOI: 10.1007/978-94-007-2762-5, © The Author(s) 2012